Table of Contents

by Steve Glass, PhD, FASCM
Brian Hatzel, PhD, AT, ATC
Rick Albrecht, PhD

Kinesiology For Dummies®

Published by: **John Wiley & Sons, Inc.,** 111 River Street, Hoboken, NJ 07030-5774, www.wiley.com

Copyright © 2014 by John Wiley & Sons, Inc., Hoboken, New Jersey

Published simultaneously in Canada

For general information on our other products and services, please contact our Customer Care Department within the U.S. at 877-762-2974, outside the U.S. at 317-572-3993, or fax 317-572-4002. For technical support, please visit www.wiley.com/techsupport.

Wiley publishes in a variety of print and electronic formats and by print-on-demand. Some material included with standard print versions of this book may not be included in e-books or in print-on-demand. If this book refers to media such as a CD or DVD that is not included in the version you purchased, you may download this material at http://booksupport.wiley.com. For more information about Wiley products, visit www.wiley.com.

Library of Congress Control Number is available upon request

ISBN 978-1-118-54923-0 (pbk); ISBN 978-1-118-54924-7 (ebk); ISBN 978-1-118-54925-4 (ebk); ISBN 978-1-118-54926-1 (ebk)

Manufactured in the United States of America

10 9 8 7 6 5 4 3 2 1

Contents at a Glance

Introduction

. . . .

*L*ife is movement. Starting from the smallest living cells to the most dynamic athletic skill, all aspects of the human body are in a constant state of movement. Kinesiology is, literally, the science of movement. As a student of kinesiology, you'll study how the body initiates and controls movement, starting with the brain and involving all the different body systems.

Part of the beauty of movement is its complexity. Successful movement happens only when the different systems — the cardiovascular system, the neuromuscular system, and energy-producing system — work together in an organized way. What's so amazing about movement is that it requires the coordination of so many systems, all interacting in a constantly changing relationship — one that most people take completely for granted.

In addition to studying the nuts and bolts of movement itself, kinesiology also examines how body systems interact during various forms of training and how to use that knowledge to enhance performance, avoid or overcome injury, and promote physical fitness in individuals and whole populations.

You may be a coach, an aspiring professional in kinesiology or its many related fields, or an athlete who wants to know more about his or her body. This book is for you! It contains the primary principles of kinesiology, the mixture of subdisciplines (sport psychology, biomechanics, and exercise science, for example) that you'll be exposed to, and enough information to set you on a path of understanding the complex human systems involved with movement. In short, this book contains key information that will form a solid foundation for you on your journey of understanding.

About This Book

This book contains the primary principles of the field of kinesiology. Kinesiology is a vast field, and textbooks within even the subdisciplines themselves contain so much information that you can get lost in the details. This book, on the other hand, covers the major concepts you need to know — the key aspects of kinesiology across all the major systems of the body, the primary subdisciplines and so on — in a much easier-to-read-and-understand format than you'll find in other textbooks in the field.

Kinesiology For Dummies is an excellent introductory text to the entire field of kinesiology. Here, we show you the forest rather than force you to focus too much on the trees. To make information easy to understand, we use the following conventions:

- We gradually introduce you to the jargon you'll hear as a kinesiology student to help ease you into the complexity of the information. With the basic concepts under your belt, you'll be able to confidently pursue greater depth of learning that will come as you get further into the subject.

- We've sprinkled sidebars and paragraphs accompanied by Technical Stuff icons throughout the book. We added these to give you a glimpse at some of the more detailed information that is out there. You can skip these tidbits if you like, but if you're hungry for more information or deeper understanding, these discussion can help point the way.

- Within this book, you may note that some web addresses break across two lines of text. If you're reading this book in print and want to visit one of these web pages, simply key in the web address exactly as it's noted in the text, pretending as though the line break doesn't exist. If you're reading this as an e-book, you've got it easy — just click the web address to be taken directly to the web page.

Foolish Assumptions

Although we assume you don't have a substantial background in kinesiology field, we do assume that you have some knowledge of anatomy and basic physiology — things like the anatomy of muscle, the skeletal system, the heart and circulatory system, as well as the basic physiology of how these systems function within the body (how the heart beats, blood moves, muscles contract, and so on). We offer a very basic overview of these topics, but if they're completely unfamiliar to you, consider skimming through *Anatomy & Physiology For Dummies,* 2nd Edition, by Maggie Norris and Donna Rae Siegfried (John Wiley & Sons). It's an excellent way to build a foundation of the body at rest, before you dive into a book about the body in motion.

Here are a few other assumptions we've made about you:

- **If you're an avid exerciser,** you may be reading this book for your personal use, because you want to know more about how the body works. We assume you want the main points, the big picture, and useful information that will help you in your training.

✔ **If you're a budding kinesiology student,** this book is an excellent way to get a broad view and some key information about kinesiology and its many subdisciplines. Reading this text as part of an introductory course in the movement sciences would be a very good first step in your training.

✔ **If you're a coach or personal trainer,** you have a background in one or more aspects of kinesiology, but you're trying to broaden the scope of your knowledge. Consider this book a refresher in material you may have learned already and an introduction to topics you may not have previously been exposed to.

Icons Used in This Book

You'll notice some images along the page margins. These icons clue you in on particular types of information within the book:

This icon points you in the direction of understanding. Sometimes just a simple statement can make you think, "Aha! Now I get it!"

This icon summarizes and reiterates important information that you need to know. Keep these tidbits filed away for later.

An important aspect of studying movement is being able to recognize when an activity or situation increases the likelihood of an injury. We highlight these situations with this icon. When you see it, pay close attention so that you can avoid potentially dangerous situations.

This icon highlights information that we just had to share! We consider these points important enough to include but a bit too technical or slightly beyond the scope of the text. You can read these for added information or skip them.

Beyond the Book

Kinesiology requires quite a bit of background information. Check out these other *For Dummies* books, all published by John Wiley & Sons, Inc.; they're excellent resources for the additional information that can help you if you decide to pursue kinesiology as a course of study:

✔ *Anatomy & Physiology For Dummies,* 2nd Edition, by Maggie Norris and Donna Rae Siegfried

✔ *Biology For Dummies,* 2nd Edition, by Rene Fester Kratz and Donna Rae Siegfried

> ✔ *Chemistry For Dummies,* 2nd Edition, by John Moore
> ✔ *Psychology For Dummies,* by Adam Cash
> ✔ *Physics I For Dummies,* 2nd Edition, by Steven Holzner

In addition to the material in the print or e-book you're reading right now, this product also comes with some access-anywhere goodies on the web. For quick, anywhere reminders of key kinesiology topics, check out the free Cheat Sheet at www.dummies.com/cheatsheet/kinesiology. There you'll find information on how to strengthen the cardiovascular system, how the body produces the fuel it needs to sustain activity, ways to spur muscle growth, the steps to follow to perform motion analysis, and more. We've also provided lots of bonus material at www.dummies.com/extras/kinesiology that goes beyond the content in both the print and e-books.

Where to Go from Here

This book is designed so that you can jump in anywhere. You don't need to begin at the beginning. Do you see a chapter that interests you? Start there! If you're not sure where to start, head to the table of contents or the index to find specific topics that may interest you.

Each chapter stands alone. You don't need to read them in order. The key thing to remember is that this book is designed so that you can jump in anywhere, get the info you need, and jump back out. Jumping, as you'll soon discover, is a great way to approach both this book and the study of movement.

Part I
Getting Started with Kinesiology

getting started
with
kinesiology

In this part...

- ✔ Discover the basics of kinesiology, like how to take a systems approach to the control of the body and the connection between mind and body related to movement and physical fitness

- ✔ Get up to speed on the disciplines you'll be exposed to as a student of kinesiology

- ✔ Trace the history of kinesiology, from its ancient beginnings to the advanced sport and exercise programs of today

Chapter 1

Introducing Kinesiology: The Science of Movement

*T*he human body was made to move. Your health depends on it, your survival is supported through it, and your ability to engage and interact with the world requires it. Kinesiology is the science behind movement, and it examines movement in a variety of areas, ranging from health and physiology to biomechanics and sport performance.

Because the human body is complex, the study of movement is complex as well. In this chapter, we offer a quick overview of the science, the field, and the options available to you as a student — official or not — of kinesiology.

Getting Familiar with Key Areas of Study

Kinein is a Greek word meaning "to move," and the study of movement is the foundation of the wide-ranging field of kinesiology. Kinesiology covers a broad array of disciplines that examine the human body at rest, during motion, and as it adapts and changes as a result of motion.

Forming the foundation for kinesiology

Before you can understand how the body moves and adapts to movement, you must understand the human body at rest. These basics — knowing important biological processes, explaining the function of the body's structural components and its systems, knowing the chemical reactions that occur in the body, being familiar with principles governing matter in motion, and so on — give you a working knowledge of the human body and how it works.

Here's a quick rundown of the subjects you need to know *before* you get into kinesiology, arranged in a way to give you a glimpse of how the body works:

- **Biology:** Learning about living organisms and what make them tick sets you on the right path. Biology helps you understand the structure and function of cells, their growth and development, and how they come together to form complex life forms.

- **Anatomy:** When you understand how organisms function at the level of the cell, you can then begin to understand how humans (and animals) are constructed. Understanding anatomy gives you the blueprint of a species. Anatomical study ranges from the structure of the very small (cells and tissues) to the very large (the hip-bone-connected-to-the-thigh-bone kind of info).

    If you want to learn how to train someone to increase muscle growth or bone strength, you really need to know how the muscles and bones are constructed!

- **Physiology:** With a firm understanding of cellular processes (biology) and how the body is put together (anatomy), you can start to examine how cells, tissues, and organs work together in a living body. *Physiology* examines the functions of the living tissues of the body. Whereas anatomy teaches you how the heart is constructed, physiology shows you how it works in relation to the lungs and the muscles and reveals its purpose throughout the body. By studying human physiology, you begin to see that the different structures of the body are designed for specific functions that, altogether, keep the entire body functioning.

- **Chemistry:** Humans are made of matter and require energy to live. Because the body is constructed of atoms, and energy is exchanged through the interaction of various atoms, molecules, and enzymes, you need a basic understanding of chemistry. This knowledge helps you understand what goes on in the body during exercise. After you know the basics of chemistry, you can then focus more closely on the chemistry of the human body.

- **Biochemistry:** Biochemistry gives you more in-depth understanding about how the body makes energy from the food eaten and how it uses that energy to keep the cells alive.

- ✔ **Physics:** Bodies are always in motion, even when they seem to be sitting still. Therefore, understanding matter in motion — the realm of *physics* — is essential to the study of kinesiology. Physics helps you understand the relationship between energy and force, levers (like joints!), center of gravity, and acceleration.

- ✔ **Psychology:** You can't fully understand movement unless you also understand the brain! Not only do you need to know the anatomy and basic physiology of the functioning areas of the brain, but you also must have an understanding about how the brain can adapt, learn, and develop new ways of moving the body. This area also delves into how emotions influence the body and behaviors.

Getting serious: Embarking into the fields specific to kinesiology

Sometimes the hardest part of starting a career in kinesiology is deciding which field to focus on! Your interest may gravitate toward the microscopic: the actions of cells and organ systems and how they function during movement. Maybe you'd prefer to focus on the way the body performs movements and generates forces, or how the body heals through physical training. There is a field for all interests within the study of kinesiology. Take a look at some of the primary fields listed here.

Exercise physiology

Exercise physiology is all about the body in motion. Understanding how the systems of the body (like muscular and cardiovascular) behave during exercise and how they adapt as a result of exercise training is a major part of exercise physiology. For detailed information on exercise physiology, head to the chapters in Part II.

Exercise is used as a tool to change the body as well as to better understand how the body functions. For this reason, exercise physiology is a key component of the many careers and fields that use exercise as a way to improve the body. You can discover a number of these fields in Chapter 18.

Biomechanics

Movement involves forces, levers, balance, and accelerations. Starting with a foundation of mathematics and physics, biomechanists study the mechanics of movement. Movements can be as simple as lifting a weight or as complex as walking (gait) or doing a high jump. Biomechanics uses technologies that can measure forces *(force platforms)* and the activation of muscles *(electromyography),* and it often uses video to analyze all the aspects of body movement. Part III delves deeply into the biomechanics of movement.

Rehabilitation therapy

Injuries can happen for a variety of reasons: perhaps from a movement that isn't performed correctly (you lift something wrong, for example), an accident (you fall on an arm), or some underlying health issue (a problem exists with your heart or lungs, for example).

Understanding how the body heals and the interaction between exercise and the healing process is an area of study that spans a number of career fields. These fields often combine medical knowledge with exercise physiology, biomechanics, and even sport psychology. Studies for this field may focus on cardiac rehabilitation, physical therapy, respiratory therapy, occupational therapy, and therapeutic recreation. Parts II and III help contribute knowledge to rehabilitation of the body.

Sport and exercise psychology

After the body has been trained for an activity, the mind becomes the most important aspect of performance. Mood, behavior, and confidence all influence performance, for better or for worse. This area of study seeks to answer questions like, "How do athletes control the stress of a competition and still do their best?" and "How can an athlete be 'in the zone' one day and then perform terribly the next?"

Sport and exercise psychology studies human behavior and the mind and applies that knowledge to determine how best to train athletes to get the most out of their performance. In Chapter 13, we take a close look at the impact of motivation (or lack thereof) on performance and offer suggestions on how to get and stay committed to a physically active lifestyle.

Strength and conditioning

Athletes' bodies can perform at their best only if they have been properly conditioned for the activity. Because movement requires conditioning the muscular and cardiovascular systems, as well as training the body to hold off fatigue, studying strength and conditioning gives you a deep understanding of how exercise changes the body. You also learn how to apply training principles that are specifically designed to improve performance in a sport. Parts II and III cover aspects of conditioning related not only to the muscles (like Chapter 10) but to the other systems of the body that are essential for peak performance.

Sports and athletics

With expertise in the skills required by a particular sport and with full understanding of all aspects of human movement, coaches serve as teachers and mentors to developing athletes. Because they perform a number of roles for the athlete, coaches need to understand the principles of how exercise can be used for conditioning, and they must know how to effectively apply the concepts of motivation and behavior change. Effective coaches also study the foundations of sport within the culture and discover strategies for motivating young athletes to perform at their best within a competitive and stressful environment.

Fitness and wellness

Cardiovascular disease and cancer are the leading killers of men and women in America. Research has shown a strong link between these conditions and physical inactivity and poor nutrition. Fitness and wellness professionals use exercise and physical movement as part of a comprehensive approach to reduce the incidence of cancer and heart disease. Exercise, body fat reduction, and dietary improvements go a long way toward putting people on a path to health. Chapter 17 delves into the link between physical inactivity and health problems related to obesity.

Understanding the Many Systems That Make Up the Human Body

Single-cell organisms have it so easy! Everything they need is contained in one cell. All their biological processes (eating, generating energy, moving, "thinking," and reproducing) have to be carried out within their single cell, and their range of interaction with the environment is quite limited. Human bodies, on the other hand, are able to adapt and interact with each other and the environment. To function at such a high level, the human body is *much* more complicated. Structurally, it has multiple levels (cells, tissues, organs, and organ systems) that build on each other and that must all function in a coordinated way to maintain the health of the organism — you.

As a student of kinesiology, you'll be introduced to the following systems. Kinesiology helps you understand how these systems interact and change as a result of movement and exercise training:

✔ **The brain and nervous system:** The brain and the neurons that make up the brain function as a central processing center where all the information about your body and your environment can be interpreted. The other systems of your body communicate with each other through the nervous system, enabling you to see, hear, move, and interact with your surroundings. This system constantly adjusts and adapts to your movements and your environment. To find out about the nervous system, head to Chapter 3. Chapter 6 explores how your body adapts to different environments.

✔ **The circulatory system:** Humans need continual sustenance to survive, and the circulatory system is the primary highway over which nutrients like glucose, fatty acids, oxygen, and hormones travel. The arteries transfer nutrient-rich blood to your tissues, and thin capillaries create easy access to the tissues. Your veins help guide the nutrient-depleted blood back to the heart and lungs for a refresher. The circulatory system changes its flow during times of stress or exercise. Chapter 5 covers the key functions of the circulatory system related to oxygen and nutrient transfer.

✔ **The cardiorespiratory system:** To keep a constant flow of nutrients coming to your tissue and to keep wastes moving out, the body needs a pump and a fueling station. Fortunately, it has both: the heart and the lungs. The heart keeps blood moving, and the lungs serve as the station where oxygen-depleted blood fills up again. Every time a *ventricle* (a chamber in the heart) contracts, its dual chambers either push blood to the lungs to pick up more oxygen (right ventricle) or push oxygen-rich blood to the entire body (left ventricle). Exercise can help train this pump to do more work, push more blood, and get you in shape.

✔ **The skeletal system:** The human body is about 70 percent water, and most of the tissue in it is made up of some pretty soft stuff. Without a frame to mount the soft, squishy bits on, we'd all be a big blob of humanity! The skeletal system provides a rigid framework that allows you to move about and see the world. Strong bones, constructed with plenty of calcium, mean a strong frame. Functioning joints enable you to move with little effort. When this system begins to weaken (and lose calcium), mobility really drops. You can read about the skeletal system and joints in Chapters 8 and 9.

✔ **The muscular system:** Movement wouldn't be possible without something to produce force. In the body, those "force producers" are your muscles. Muscles provide the horsepower you need to move your body and interact with your world. They're also very adaptable. If you make them do a lot of work, they grow stronger. If you let them sit around and do nothing, they shrink! Strong muscles play a role in good health and quality of life. Head to Chapters 7 and 10 to find out about movement in general and the muscular system in particular, and Chapter 11 to delve into motion analysis.

✔ **The endocrine system:** Although the brain can control many of the functions of the body through the nervous system, other controls require chemical stimuli. Glucose (the sugar your body uses for energy), for example, can't get into the cell unless the pancreas secretes insulin to help create a pathway into the cell. The endocrine system involves a number of organs and glands that secrete chemicals that bind to receptors both inside and outside cells to essentially open and close cell doors, either letting in or blocking out these chemicals. Sometimes the release of hormones can cause a fast response (insulin helping to drop blood glucose levels, for example); other times, the release of hormones may cause changes that occur slowly over time (thyroid hormones can slowly make changes in your resting metabolic rate, for example).

Check out *Anatomy & Physiology For Dummies,* 2nd Edition (Maggie Norris and Donna Rae Siegfried), or *Biology For Dummies,* 2nd Edition (Rene Fester Kratz and Donna Rae Siegfried), both published by Wiley, for complete discussions on the endocrine system and the role of hormones.

Examining Movement from Many Angles

Chances are that, when you hear the term *movement,* you have your own idea of what it means and how people use it in their lives. But you can think about movement and the connections between it and the world in more ways than you probably imagine. In the following sections, we outline the many ways movement can be examined.

Studying the biomechanical basis of movement

When you throw a ball, clear a hurdle, or balance on a beam, you probably focus on the result (were you successful?) or the "feel" of the movement (the power of your release, for example, or the steadiness of your stance). Others, however, examine movement from the standpoint of the forces and accelerations that are created or that result from the activity. *Biomechanists* study these aspects of movement, using the tools of physics, math, and kinesiology, to answer questions like the following:

✔ **How is movement impacted by changes in the center of gravity?** Does changing the position of the arms and legs, for example, impact how someone jumps over a high jump bar or executes a gymnastic move?

✔ **What forces and velocities exist in vertical, horizontal, and rotational dimensions?** By knowing the forces, you may change how a spin and rotation are completed in a high dive maneuver.

✔ **How is balance maintained and lost?** Do older people fall because their muscles are too weak to handle a change in direction, or is the falling due to a delay in the muscle's ability to generate the required force?

✔ **What are the ergonomics of different movements (like the forces involved with typing, sitting, and repetitive work)?** Will sitting on a ball instead of a chair, for example, actually help improve posture and reduce low back pain?

> ✔ **What is the most efficient movement (running stride, throwing motion, or jumping technique, for example), based on an analysis of the action itself?** For example, can you make a runner faster simply by changing her running technique?
>
> ✔ **How can the principles of movement be used to prevent injury?** Do ways exist to land from a jump that can reduce forces on the knee and prevent tearing a ligament? Maybe a throwing motion can be changed so that the ligaments in the shoulder can bear the forces more easily.

Focusing on the health-enhancement aspects of movement

The human body is meant to move. A body at rest begins to wither away *(atrophy)* and lose muscle mass, bone density, and even heart size. Movement can be a tool to help the body's systems function at a more optimal level. In fact, regular movement can produce the following beneficial results:

✔ Reduced blood pressure and stronger heart

✔ Increased bone density

✔ Improved blood cholesterols

✔ Stronger immune system

✔ Reduced incidence of cancer and heart disease

✔ Reduced stress, anxiety, and depression

Movement is a key component of physical health, as well as mental health. Chapter 13 examines how to set and achieve exercise goals to improve your physical health, and Chapter 14 delves into the connection between physical activity and mood.

Looking at the aesthetics of movement

The human mind has an interesting gift: enabling people to vicariously experience what they witness others doing. If you've ever cringed when someone falls or felt a rush when you see someone execute a flawless performance, you've experienced this sensation yourself. This capacity is probably one of the reasons for the worldwide popularity of sport and dance: We draw pleasure from watching movement that displays artistry, especially when it involves balance, coordination, and flawless technique.

Pick any activity, and you'll find spectators that draw pleasure from watching skilled performers: ballet (the pliés, pirouettes, and jumps), ballroom dancing (the fluid movements, spins, and complex steps), figure skating (the synchronicity between pairs skaters and the jumps), martial arts (the powerful yet beautiful kicks, the speed and grace of the movements), and so on.

Now imagine the control it takes to execute such movements, especially at a world-class level. Almost everyone can throw a ball, for example, but fewer can throw it with power. Even a smaller number can throw it with power *and* control. And only a select few with the power, control, speed, and consistency that make a world-class pitcher. The differences between the person who can throw reasonably well and one who can throw a no-hitter in a World Series game are due to many factors, including training, physiology, motivation, and more. Kinesiology helps shed light on these factors in an effort both to understand movement in individuals and to draw conclusions that can help anyone enhance performance or overcome mobility challenges.

Uncovering cultural influences on movement

The culture in the United States has a bit of a split personality when it comes to movement. In one sense, the U.S. has a highly motivated sport culture, where kids and adults train continually to improve their performance and attain peak physical condition. Traveling sport teams and elite squads of competitors give up other parts of their lives just to train, for example.

On the other hand, over 60 percent of the U.S. population is inactive and overweight or obese, a situation that has led to increases in diabetes and other chronic ailments, many of which are life-threatening and all of which take a large portion of the blame for escalating healthcare costs.

In other cultures, activity on a daily basis (walking or biking, for example) is a way of life and results in lower rates of diabetes, heart disease, and the other chronic conditions that plague Americans and, increasingly, the populations of other Western countries. The U.S.'s love-hate relationship with movement continues to evolve as people work to find a balance in their lives. Head to Chapter 12 for a detailed look at the cultural and social aspects of movement.

Determining Whether Kinesiology Is the Field for You

Because movement is an inherent part of life, the science of movement, kinesiology, is an inherently important field. Kinesiology techniques and areas of study are used by medical professionals, athletic departments, sports organizations, corporations, and many other industries to enhance performance, improve health, overcome mobility challenges, and more — all by changing the way people move.

To help you determine whether this field is the right one for you, ask yourself these questions:

✔ **Do I enjoy movement?** Many people that enter the field of kinesiology are avid exercisers, athletes, or people who just like the science of movement. In many cases, the best professionals are those who "practice what they preach" and are able to lead their clients by example. If you have always enjoyed movement, kinesiology may be for you!

✔ **Do I like helping people?** In almost all the fields within kinesiology, you use techniques to analyze and improve the movement or health of other individuals. As a result, you are in the people business! If you like helping others, especially helping others improve themselves, kinesiology may be a good fit for you. Your day may be filled with activity and not much time behind a desk, and your interpersonal skills will be a big plus as you work with people from all walks of life and in all different conditions.

✔ **Do I want to know about the science behind the movement?** If you just like activity but aren't interested in or don't enjoy learning about the science behind the movement, then you may struggle in this field. Because human movement is so complex, kinesiology is based in anatomy, biology, physiology, psychology, physics, and so on. Conversely, if you like the sciences and can't wait to apply them to human movement, then you're in for a real treat because, as a kinesiology student, you'll be immersed in all of them!

✔ **Do I want to use movement as a way to help improve the human condition?** Exercise, physical activity, and movement are tools to change the human body. Kinesiologists use these tools to help individuals heal or improve their condition. Exercise is a medicine, and kinesiology shows you how to use it effectively to help the individual.

Kinesiology provides the foundation for a wide range of careers that use movement, exercise, and rehabilitative therapies to help improve the body. As you build your knowledge of the systems of the body and begin to recognize all the ways kinesiology can be applied, you'll see the endless possibilities. Chapter 18 outlines ten careers for the budding kinesiologist. Take a gander — and good luck in your studies!

Chapter 2

The History and Philosophy of Sport and Exercise

*E*arly humans had to expend huge amounts of energy just to stay alive. They burned thousands of calories a day hunting, fishing, scavenging, and foraging for any and all kinds of food. And they no doubt consumed nearly as many calories avoiding being prey themselves. Anyone too unfit to outrun or outfight lions, tigers, and bears (and a bunch of other nasty beasties) was probably not going to be around long enough to help create the next generation.

Even after banding together to collect food and for protection from predators, they continued solving many of their problems with brute physical force. Wars were fought on a regular basis, and to keep their warriors in tip-top physical condition for the next battle, societies developed peacetime games and spectacles patterned after warlike activities. Although these games evolved over time, the remnants of these militaristic origins can still be seen in many of today's sports and exercise activities.

From these early beginnings, the science of sport and exercise medicine was born. In this chapter, we take you on a tour of the milestones between the earliest efforts to understand and improve performance and what kinesiologists do today.

Chasing Down the Art of Prehistoric Movement

Obviously, the term *prehistoric* means "before history." With no written historical record, no one knows for sure exactly how far back people started engaging in strenuous physical activity. What researchers do know, however, is that, in those earliest of days, just living from one day to the next was a matter of doing whatever was required to eat without being eaten.

Even though no written records exist that tell researchers about the physical strength and stamina of these people, clues do exist in their artwork — the hundreds of paintings and drawings left on cave walls around the world. These pictures depict the things that mattered most in their lives. In addition to painting elaborate pictures of the deer, birds, and bison they hunted for food and fur, these people also drew pictures of themselves engaging in physical survival skills like running, wrestling, swimming, throwing spears, and shooting arrows.

Although it's doubtful that prehistoric humans engaged in physical activity just for pleasure or relaxation, you can easily imagine that young kids who were not quite old enough to join in the hunt underwent some form of training in the physical skills they would later need when they became responsible for providing for themselves and the rest of the clan. In any event, you can be certain that your ancestors who lived tens of thousands of years ago recognized the important health benefits associated with being bigger, stronger, and faster than anyone or anything else in their environment.

Contemplating Eastern Philosophies and Fitness

Long before the Greeks, ancient Eastern philosophies such as Confucianism, Taoism, Buddhism, and Hinduism embraced the idea of regular exercise to promote health and fitness. The lasting impact of these early forms of exercise can still be seen thousands of years later in exercise and fitness centers around the world, where classes in kung fu, tai chi, and yoga are among the most popular classes offered.

Stressing the importance of physical activity: Chinese philosophies

Most ancient Chinese philosophies stressed the important role physical activity plays in health and wellness. It was widely understood that exercise was essential in preventing many diseases and disorders, leading to the creation of two systematic exercise programs that are still practiced, albeit it in slightly different forms, today:

- ✔ **Tao yin:** Tao yin was originally practiced by followers of the Taoist philosophy, which held that physical exercise was a way to achieve better health and to spiritually align oneself with the forces of the universe (known as *chi*). A series of Taoist exercises called *tao yin* are thought to be the basis of the popular form of martial arts we now call *tai chi*.

- ✔ **Kung fu (or *cong fu*):** Another form of Chinese exercise that predates much of what was done in Greece, kung fu is a series of stances and movements associated with Confucianism. The exercises were seen as a way of practicing self-discipline that would better the individual and society.

Focusing on the spiritual: Buddhist and Hindu philosophies

In India, leaders of the Buddhist and Hindu philosophies weren't all that keen on overemphasizing the body. They thought their followers should be more concerned with spiritual rather than physical wellbeing. As a result, they didn't feel completely comfortable with either the Taoist's tao yin or the Confucian's kung fu.

Instead they developed something that was more in line with their philosophical beliefs, which focused on mental and physical stillness and peace — and something that has become one of the hottest forms of exercise in the world today — *yoga!*

Combining Sport and Science: Greeks and Gladiators

Although followers of the Eastern philosophies had long understood the direct connection between physical activity and health, the Greeks and Romans were the ones who made a particular type of physical activity — sport — a fundamental part of their life philosophy. To the Greeks, sport had great religious significance. The Romans, on the other hand, developed a very different view of sport. To them, sport was important for its entertainment value. They were generally more concerned with the sights, sounds, and spectacles surrounding sport — and the more outlandish and over-the-top the event, the better.

The Greco-Roman period lasted roughly 1,300 years (from about 800 BCE to 476 AD), and it made sport, in one form or another, a central part of almost everyone's daily life. With this new emphasis on sport came a keen interest in the underlying science of performance, which quickly led to a far better understanding of exactly how the human body works and what could be done to push it to its limits.

Games, gods, and glory in ancient Greece

In 776 BCE, some of the first athletic competitions in history were held at Olympia in Greece. These games took place in the shadow of Mount Olympus because that's where the Greeks believed their gods lived. Although the games at Olympus (held in honor of the father of all gods, Zeus) may be the only ones still remembered today, religious-based athletic festivals were quite the rage throughout the Greek city-states.

Although, like today, winners of these first Olympic games were well-compensated, the Greek games were first and foremost a way to honor their gods. Some of the main athletic events and religious festivals took place in Delphi (in honor of Apollo, the god of light), Nemea (also held for Zeus), and Isthmis (honoring Poseidon, the god of the sea). Today, those who are tempted to criticize athletics for becoming something akin to religion would be wise to remember that, in ancient Greece, the very first athletic completions were virtually nothing but religious celebrations!

The "Big Daddies" of exercise science

Because athletics was interwoven with Greek religion and philosophy, it soon became an integral part of the entire society and included something the Greeks were really big on: scientific investigation. In the following sections, we introduce you to four ancient physicians and philosophers who made significant contributions to the field.

Herodicus: The father of sports medicine

Around 480 BCE, a physician/sports teacher (coach) by the name of Herodicus was the first Western physician to combine sports, exercise, and medicine. He was known far and wide for recommending that his patients eat a healthy diet, exercise vigorously every day, and take frequent therapeutic massages. Still, Herodicus may be best known for the ideas he passed on to his star pupil, Hippocrates, the father of Western medicine. (You can read more about Hippocrates in the next section.)

One indication of the impact that Herodicus's thinking has had on the field of sports medicine today is that, nearly 2,500 years after his death, one of the most prestigious orthopedic sports medicine societies — the Herodicus Society of the American Orthopaedic Society for Sports Medicine — still bears his name.

Hippocrates: The father of Western medicine

Hippocrates (460–370 BCE) followed in the footsteps of his teacher, Herodicus, by placing a tremendous emphasis on the health benefits of diet, exercise, and overall fitness. In fact, a quote attributed to Hippocrates is something most kinesiologists and fitness/wellness specialists would be proud to put on the back of their business cards today:

> If we could give every individual the right amount of nourishment and exercise, we would have found the safest way to health.

Not bad for a couple thousand years ago.

Aristotle: The father of kinesiology

Aristotle (384–322 BCE) was the first person to study and describe general body movements and the forces required to move various parts of the body. He also wrote the first book on the subject of biomechanics and is the one who came up with the term *kinesiology* in the first place: The Greek word *kine* means "to move," and *ology* means "to study or discuss," so Aristotle was simply saying that kinesiologists study human movement.

Although people generally think of Aristotle as a philosopher — after all, he was one of Plato's students — he actually knew pretty much everything about pretty much everything.

Archimedes: The father of mathematical physics

In the process of inventing all sorts of military machines, Archimedes (287–212 BCE) discovered many of the basic laws of physics that would later be used to better understand and describe human movement. Two of his laws in particular made huge contributions to the field of kinesiology:

✔ **The Law of the Lever:** Archimedes figured out how a long rigid object (like a bone) can be put over a pivot point (like a joint) to move large objects (like an arm or leg) with relative ease. In fact, he's famous for saying, "Give me a long enough lever and a place to stand, and I will move the world." Sir Isaac Newton later used this law to come up with Newtonian physics. To get an idea of just how important the Law of the Lever (and Newtonian physics) is to the current understanding of biomechanics, just take a look at the discussion of human motion in Chapter 7.

✔ **The Law of Buoyancy:** This law reflects Archimedes's understanding of water displacement — which basically means that objects will sink or float based on their density relative to the density of the fluid in which they are immersed. This law remains the basis of athletic performance in water sports and aquatics today, and it's the reason researchers and technicians throw people into tanks of water to see how much fat they have on their bodies (see Chapter 12 for more information about hydrostatic body composition testing).

Making spectacles of themselves: The Romans

The Romans had their own ideas of what sport and exercise were all about. Unlike the Greeks, who saw individual athletic performance and excellence as a way to honor their gods, the Romans were far more interested in spectacle and entertainment. Basically, these guys took their X Games to a whole new level.

This is war (or at least something very much like it)

To the ancient Romans, the more blood and guts spilled during athletic competitions, the better. Thousands of citizens would jam into the Colosseum or the Circus Maximus just to watch people and animals get ripped to bits for nothing more than entertainment. Gladiators fought to the death; huge animal hunts were staged; and bloody re-creations of battles took place for the enjoyment of the crowds. These early fans attended not to appreciate the finer points of athleticism but to witness death and dismemberment. As a result, no one was particularly interested in scientifically studying human performance. There was one notable exception, however: Galen, whom you can read about in the next section.

Galen: The physician to the gladiators

Galen (129–200 AD) was a Roman physician who conditioned and trained gladiators. As he patched up his wounded and dying patients, Galen made some interesting observations about the way human anatomy, physiology, and neurology actually worked. For example, he was the first person to see that there were two "types" of blood: one dark (venous) and the other bright (arterial). He was also the first person to identify two separate nerve pathways: one for receiving information (sensory) and the other for sending messages out to the muscles (motor).

Interestingly enough (given what they were doing to actual living people in the Colosseum), the Romans had strict laws against dissecting human cadavers. In fact, it was a lot easier to see the inner workings of a human being from a front row seat at the Circus Maximus than to stand around an autopsy table. Prior to his work with the gladiators, Galen, who was considered the top physician at the time, was limited to dissecting living and dead animals. You could say that getting hired to work with gladiators gave him his big break.

Watching the Rebirth of Kinesiology: The Renaissance

Not a lot happened between 476 AD and the 15th century, a period known as the Dark Ages. But then the flames of scientific knowledge slowly started to be rekindled, one small candle at a time. (Read the sidebar "Hey! Who turned out the lights?" for details on why scientific inquiry came to an almost complete halt during this period.)

Although plenty of people during the Renaissance (which means "rebirth") made valuable contributions to what scientists now know about the human body and how it works, four pillars of art and science did more than all the others to get the study of kinesiology back on track and headed in the right direction: Leonardo da Vinci, Galileo Galilei, William Harvey, and Isaac Newton.

Leonardo da Vinci: A renaissance man among renaissance men

The term *renaissance man* has come to signify a person with many interests and exceptional expertise in many areas. If anyone ever epitomized that description it was the Italian inventor, architect, musician, painter, sculptor, mathematician, engineer, and — most important to those studying kinesiology — anatomist and artist, Leonardo da Vinci (1452–1519 AD).

Hey! Who turned out the lights?

As a new religious philosophy — Christianity — started to take hold across the Roman Empire, peoples' appetites for killing and maiming other humans just for sport waned. Interest in sport, science, medicine — and almost everything else for that matter — came to a screeching halt with the fall of the Roman Empire around 476 AD. For more than a thousand years, from the end of the Roman Empire to the Renaissance and the Western explorers' discovery of the New World around 1500 AD almost nothing happened to further the understanding of the workings of the human body or contribute in any way to what kinesiologists do today. This thousand-year dead zone of intellectual and scientific inquiry is often and quite appropriately referred to as the Dark Ages.

Although many people first think of da Vinci as the painter of some of the most famous art masterpieces in all of history (like *Mona Lisa*, *The Last Supper*, and the *Vitruvian Man* drawing), in medical circles, he's best known for using his artistic genius to create some of the most accurate and detailed illustrations of the human body before the invention of computer-generated imagery. Da Vinci's almost photo-quality drawings of the muscles, tendons, ligaments, and bones are so precise and clear they still appear in anatomy textbooks throughout the world.

Galileo Galilei: Measuring heavenly bodies (and ours) in motion

Like his Italian countryman da Vinci, Galileo Galilei (1554–1642 AD) was also a true renaissance man, but he tended to stick more with the sciences. Also like da Vinci, Galileo is probably best known for his accomplishments that are totally unrelated to kinesiology. Because of his work as an astronomer, physicist, mathematician, and philosopher, he is considered the father of physics and even the father of modern science.

Galileo's major contribution to kinesiology came in 1638 when he wrote his last book, *Discourses and Mathematical Demonstrations Relating to Two New Sciences*. The two "new" sciences he wrote about would today be called kinematics and strength of materials. Both of these sciences have to do with kinesiology, but *kinematics* (the geometry of motion) is so important that you'll find it discussed in several places in Chapter 7.

What happens when you tick off Church elders

Although the Dark Ages are officially thought to have ended around 1500 AD, more than a century later, the Roman Catholic Church was still going after scientists it viewed as heretics (basically anyone who disagreed with Church doctrine).

In 1638, one of the most famous scientists of all time (and a big-time contributor to kinesiology) — Galileo Galilei — got into serious trouble with Pope Urban VIII and the Roman Inquisition for agreeing with the Polish astronomer Nicolaus Copernicus (1473–1543 AD) that the sun was the center of the solar system

(a position technically called *heliocentrism*). The Church saw heliocentrism as a direct attack on God because Church doctrine taught that humans were so important to God that he must have made our home, Earth, the center of the universe (a view called *geocentrism*).

It turns out, of course, that Galileo was right all along, but that didn't keep the Pope from putting him under house arrest until Galileo's death in 1642. The good thing is, while locked up, Galileo had time to write his book *Two Sciences,* which kinesiologists find so valuable.

William Harvey: Figuring out the lifeblood of kinesiology

It's pretty hard to think about doing any strenuous physical activity without the heart pumping away like crazy, sending blood and oxygen (and a whole lot of other important stuff) to every part of the body. But believe it or not, it wasn't until 1628 that someone — an Englishman named William Harvey (1578–1657 AD) — finally figured out what the heart, the lungs, and the entire circulatory system were actually up to.

You can see exactly how much Harvey contributed to kinesiology by looking at what scientists thought circulation was all about *before* he came along. Here are a few tidbits:

- They thought the heart's job was to heat the blood, and the lungs' job was to cool it.

- They thought (courtesy of Galen; see the earlier section "Galen: The physician to the gladiators") that two entirely separate blood systems existed: arterial blood originating in the heart and venous blood originating in the liver.

- They thought that, when the arteries dilated, they were sucking in air and that, when they contracted, they were sending "vapors" out through pores in the skin.

To see what is known today and to get a good feel for all of Harvey's contributions to kinesiology, just take a quick glance at Chapter 5.

Isaac Newton: The lawman of motion

Newtonian physics, named after English physicist and mathematician Isaac Newton (1642–1727 AD), is the basis for the study of body movement. In his book *Mathematical Principles of Natural Philosophy,* published in 1687, Newton put together the three main laws of physics that govern movement.

As you may expect, because kinesiology is the study of human movement, Newton's three laws of movement are pretty important:

✔ **Inertia:** When an object is moving, it will continue to do so until an outside force is applied; when an object is at rest, it will remain at rest until an outside force is applied.

✔ **Acceleration:** The rate at which an object changes *velocity* (speed and direction) is determined by two variables: the net force acting upon the object and the mass of the object.

✔ **Action-reaction:** For every action, there is an equal and opposite reaction. In other words, when an object exerts a force onto another object, the contacted object exerts a reaction force that is in the opposite direction and of the same magnitude.

These three laws are the basis of biomechanics, an important subfield of kinesiology. To see how important biomechanics and Newton's laws are to kinesiology, head to Chapter 7.

Let the Movement Begin! Into the 19th and Early 20th Centuries

The 19th and 20th centuries provided a catalyst to what is now modern physical activity and sport. Before the 19th century, many religious sanctions on exercise existed. As those sanctions were removed and exercise was encouraged, a tremendous boom in activity occurred. Many early systems were based on individual benefits, whereas others evolved from militaristic roots.

During this time, many of today's popular sports, like basketball (1891), football (1869), and soccer (1877), had their beginnings. In addition, the first Modern Olympic Games were held in Athens in 1896.

Religion, military preparedness, and most certainly spectator interest precipitated and influenced this growth in activity. As a result, physical activity today no longer solely represents survival skills; it emphasizes team and individual accomplishment and life-long involvement, and it's supported by organized sport organizations with governing bodies (like the NCAA) and academic research.

Running through exercise and fitness philosophies

During the past two centuries, many philosophies have evolved into what is considered modern sport and physical activity. These philosophies, sometimes referred to as *systems,* were often based on and emphasized naturalistic, nationalistic, moral, and religious priorities. Often dependent on political influences or social perspectives of the time period, physical activity grew out of the needs of the both individuals and their countries.

In the following sections, we introduce you to several early physical education philosophies. Keep in mind, however, that this list isn't exhaustive. The ones we included are those that have had a significant effect on what is known today about physical education and sport.

Activity for yourself: The naturalism movement

The *naturalism movement* focused on the personal benefits of physical activity. The thinking revolved around the belief that explorations and adventures in the natural environment (nature) — like going out into the wild to hunt — prepare people physically and intellectually. They believed that these kinds of activities made them stronger and more fit.

The defining characteristic of the naturalism movement was the focus on the individual and activities that were designed to benefit them personally.

Activity for your nation: The nationalism movement

The *nationalism movement* in exercise was based on the idea that a strong nation resulted from a physically developed society made up of active people. This philosophy drove the German, Danish, and Swedish exercise systems:

✔ **In Germany:** The German gymnastics system sought to develop fitness and strength in German youth with the goal of unifying the country. The birth of the *turnplatz,* or outdoor exercise area, came out of the German program. Many years later, Adolph Spiess (1810–1858), the German gymnast and educator, was credited with developing German School Gymnastics. The emphasis with this system was to integrate gymnastics beliefs into the schools. It stressed discipline and obedience and included a variety of exercises with musical accompaniment.

✔ **In Denmark:** The Danish system was based on fitness, strength, and military competence. Known originally as *Nachtegall's curriculum,* it gained popularity and later was documented in the *Manual of Gymnastics* in 1828. This system was based largely on command-response exercises, with rigid, mass drills around a nationalistic theme, much like the German system.

✔ **In Sweden:** The Swedish movement also focused on military training, although programs included medical, educational, and aesthetic perspectives as well. Exercises had a therapeutic effect; that is, by exercising, a person could realize health benefits. The exercises were command driven, required that participants hold positions and postures, and were considered appropriate for both males and females.

These nationalist systems illustrate how political ideologies have had a significant impact on the history and development of physical activity and sport around the world. The importance of having a strong military has significantly influenced the progression of physical education philosophies and has provided individual as well as governmental benefits.

Muscular Christianity

During the mid-1800s, a movement referred to as *Muscular Christianity* took hold. The long held prohibition by various religious sects began to loosen. No longer was physical activity considered sinful.

The grounding philosophy of Muscular Christianity was that being engaged and active in sport and exercise provided a means by which one could learn and build moral values. Many proponents of this system equated physical frailty to moral ineptitude. The body was considered God's gift and the soul's temple, and the duty of Christians was to care for their bodies in service to God. Muscular Christianity greatly influenced the educational system and resulted in programs being designed to strengthen team work, honor, self-discipline, and gamesmanship.

The Sargent and Hitchcock systems

The German and Swedish systems, outlined in the earlier section "Activity for your nation: The nationalism movement," influenced the development of physical education in the United States. Also impacting U.S. physical education was the development of two early American systems — the Sargent system and the Hitchcock system:

- ✔ **The Sargent system:** This system, developed by Dudley Sargent in 1879, was based on the use of anthropometric (body) measurements to prescribe individual exercise programs for each student. Much of the program involved exercise machines, although participants were encouraged to participate in other activities such as baseball, bowling, boxing, fencing, and rowing.

- ✔ **The Hitchcock system:** Developed by Edward Hitchcock, the director of the Amherst Department of Hygiene and Physical Education, the Hitchcock system included class exercises accompanied by music. Students had class four days each week and were allowed to use a portion of their class time to practice sport skills or otherwise exercise. The Hitchcock system also used anthropometric measurements to gauge the fitness of the students and to make comparisons from year to year.

Setting academic standards for physical education

Because of the various programs and philosophies that existed in the late 1800s (and outlined in the preceding sections), in 1885, the first conference on physical training was hosted by the newly formed Association for the Advancement of Physical Education (AAPE).

The conference's purpose was to educate participants on the different systems and foster open discourse about which physical education system was best. Although no single program was found to be the best (all the systems had weaknesses), leaders from across the country learned about a number of programs, and a forum for idea exchange and advocacy was formed.

In the years following the 1885 AAPE conference, state legislators around the nation began to pass laws mandating that school curricula include physical education components. These mandates prompted existing schools to begin developing physical education programs and new schools to be formed.

With the development of set curricula for physical education programs, a more academic approach began to develop. As questions arose about the science of activity and the effects of the various programs, a movement to create a forum for discourse was created. As noted earlier, the first such forum was the AAPE; you know it today as the American Alliance for Health, Physical Education, Recreation and Dance (AAHPERD; http://www.aahperd.org).

Promoting play and amateur sport — for the leisure class

Despite the diverse perspectives of the various international systems that sprung up during the 1800s, one thing everyone acknowledged was that activity was a good thing. In the U.S., this awareness led to the following:

- ✔ **The U. S.'s playground movement:** This movement focused on providing opportunities to allow children the simple luxury of playing. Many play places for children were sponsored by local governments, which recognized the benefits that play offered in terms of socialization, health, and moral character.

- ✔ **The growth of amateur and professional sport:** Competitions grew and provided many opportunities for participation or spectating. Most of the participants in amateur and professional sports were members of the upper class, because they had the time and money to participate.

As opportunities grew, so did the disparity between the amateur and professional athletic organizations. Many amateur organizations looked down on the professional organizations and their athletes, who were thought to compete solely for the money — a focus often associated with unethical practices like cheating. Amateur organizations and their athletes, on the other hand, were perceived to exalt skill, sportsmanship, and all sorts of other good things.

Tracking Physical Ed from the Mid-20th Century to Today

Exercise and physical activity during the late 19th and early 20th century were luxuries reserved for the wealthy. Average people — both children and adults — labored hard for long hours nearly every day just to survive. They had little interest or time for what was thought of as frivolous physical activity. Meanwhile, members of the leisure class engaged in activities like bicycling, archery, croquet, tennis, hunting, and golf.

One thing changed that picture, leading, in the years to come, to physical education finally reaching the masses: the labor laws. Once the labor laws limiting the length of the workday were enacted and actually enforced, employees found themselves with more free time in which to pursue other activities and interests. Recreational activities followed.

The first modern Olympic Games

As the American amateur sport movement grew, so did the international movement. Pierre de Coubertin in 1896, following the establishment of the International Olympic Committee a few years earlier, hosted the Modern Olympic Games in Athens, Greece, the birthplace of the ancient games. This event included 12 nations, 246 athletes (all male), and 43 events. It is credited with being the start of the international Olympic movement in sport.

The Olympic Games are grounded in fair competition and sportsmanship and bring together countries from all over the world. Pierre de Coubertin fashioned the Olympic Creed — "*The most important thing in the Olympic Games is not to win but to take part, just as the most important thing in life is not the triumph but the struggle*" — placing emphasis on participation.

Since the first modern Olympics, significant growth has occurred. Today 200 nations and over 10,000 athletes (nearly half of whom are female) participate. A result of this growth can be seen in the number of national and international bodies governing sport, not to mention the multibillion-dollar television contract for broadcasting the latest contests.

A push to be more inclusive

Leading up to the 1930s and the Great Depression, physical education really began to take shape and become more inclusive. Here's a summary of some of the key changes that occurred during this time period:

- ✔ The AAPE (refer to the earlier section "Setting academic standards for physical education") began to prosper, as did many university preparatory programs.

- ✔ Nearly 30 states passed state physical education legislation mandating activity as part of school curricula. The benefits of physical education were apparent across the entire school curriculum. Physical education was considered a vital portion of the development of the student.

- ✔ Sport and recreational activities became integral parts of university life. These activities were overseen and led by faculty members at institutions like Harvard, Yale, Amherst, and Brown Universities. During this period, many intercollegiate events began to occur.

A result of the intercollegiate competition was the growth of spectating. Many folks could now be involved with physical education and sport through ways other than personal participation. The number of spectators increased tremendously for both collegiate competitions and professional sports.

Programs under fire: Examining the effects of the Great Depression

In the years leading up to the Great Depression, physical education became a mainstay of educational curricula. It was widely accepted that physical activity offered students and others myriad benefits, and there seemed little doubt that physical education and sport programs were here to stay. And then the Great Depression hit. University sport programs came under fire (which was bad), but more emphasis was placed on participation (which was good).

Decreasing number of spectators, decreasing support

University sporting programs relied heavily on paying spectators to fund their programs and facilities. As times became financially more difficult and Americans focused on making ends meet, the number of spectators at college games dropped, greatly affecting a main source of the programs' funding.

In addition, state legislatures also began to view physical education as an unnecessary expense, and state support fell. Nearly 40 percent of state physical education programs were defunded. Some states even passed legislation disallowing physical education in the schools, a complete 180-degree shift from the previous 20 years.

If you can't watch 'em, join 'em

The Depression wasn't all negative for physical education, however. True, the number of people able to pony up money for tickets dropped dramatically, hurting university sporting programs, but the dropping attendance produced one positive outcome: It placed a larger emphasis on participation rather than spectating.

Although established physical education programs were forced to change, recreation took hold and grew in their place. Participation became the key; no longer was everything dependent on competition. With this new focus on participation, many sports clubs sprang up that allowed community participation.

As more and more community sport clubs popped up, it became clear that organization and oversight were required, and many entities stepped up (or were created) to play a role. In addition to the AAPE (refer to the earlier section "Setting academic standards for physical education"), the federal government provided funding for and established programs to support public recreation. The organizations created during this period include the National Parks Service, the Forest Service, the Works Progress Administration, and the

National Youth Administration. All provided opportunities, facilities, and programs for Americans to increase their involvement in new and engaging ways.

Seeing soldiers physically unfit for duty

The Depression brought with it a much different perspective about the importance of physical education for the American public. The focus shifted from spectating to participation and the avenues to do so grew, often through sports clubs and federal programs. Yet because folks were pushed to their limits just to provide for their families, many Americans were unable to take advantage of the opportunities, and active participation in sports fell.

Despite the earlier growth in participation-based involvement (explained in the preceding section), recreational opportunities beyond spectating diminished. Such ebbs and flows through history illustrate the close relationship between physical activity and the fiscal stability of the country.

The state of affairs became strikingly apparent when a U.S. government report on the health of the nation was released after World War II. The report's key finding was that, of all the people recruited by the Selective Service Act (the draft), 30 percent failed the test evaluating physical readiness for military activity. Of those who did pass the physical examination, many more were simply inept for basic training activities. America was simply unfit for war! You can read more about this report in the next section.

Focusing on fitness again: The Eisenhower years and the Kraus-Weber test

After the war, people again began to emphasize sport, in particular *lifetime sports,* those that can be carried out throughout the lifespan: walking, jogging, hiking, swimming, and biking, for example. Despite the growth in lifetime activities, low fitness outcomes of the nation continued to be reported. It seemed that people, overall, just weren't too worried about their fitness levels — until a groundbreaking and riveting assessment of the health of American children came out in the middle 1950s.

The now widely known Kraus-Weber test results were published during the mid-1950s in the *Journal of Health, Physical Education and Recreation.* The test was a minimal muscular fitness assessment that examined abdominal strength and flexibility. Although not a thorough assessment of fitness, it did provide a comparison between European and American children. The results were chilling: Nearly 60 percent of American children failed, whereas only 9 percent of European children didn't make the cut. The results implied a weak and frail youth in America, which was a picture of the future.

In response to the reports and social concern, President Eisenhower, who was himself an avid proponent of physical activity, formed the President's Council on Youth Fitness. This group responded to the Kraus-Weber results with a recommendation that children engage in 15 minutes of vigorous activity daily.

The concern created by the Kraus-Weber report spurred the creation of various fitness-related programs, and things took a positive turn: Children improved in nearly each area that was tested. Unfortunately, the focus on regular activity eventually wore off. By the 1980s, U.S. children were once again woefully unprepared for physical activity.

Eliminating gender discrimination in education: Title IX

On June 23, 1972, President Nixon signed into law the Higher Education Amendments of 1972. The most famous portion of the amendments is Title IX, which prohibits discrimination on the basis of sex in educational institutions receiving federal aid. This legislation covers all education activities and has jurisdiction over fields of education (math and science education), dormitory facilities, healthcare, and extracurricular activities like band and other clubs.

Institutions can achieve compliance with Title IX as it relates to athletics in three ways:

- ✔ Providing athletic participation opportunities that are proportionate to the student enrollment. In other words, the opportunities for involvement need to closely resemble the gender distribution of the institution.

- ✔ Demonstrating continual expansion of athletic opportunities for the underrepresented sex.

- ✔ Accommodating the interest and ability of the underrepresented sex.

Since the implementation of Title IX, the growth in women's participation opportunities has been staggering; women have more opportunities now to participate in collegiate sport and in many more ways than ever before, up nearly 500 percent since the time the law was enacted.

Sputnik and its lasting effects

Physical fitness of American youth wasn't the only element under fire during the mid-1950s. It had become evident during World War II that America lacked technological prowess. As a result, the educational systems in place during that time came under scrutiny. Critics, many of whom were leaders in their respective fields, indicated that U.S. education in the sciences, math, and engineering was significantly lacking and disjointed. Curricula were presented in a fragmented way that rewarded memorization and mechanical computation rather than analysis. It was clear something had to change.

Many of the fears related to the shortcomings of American education became real in 1957, with the launch of Sputnik, a Russian orbiting satellite. This rather jarring wake-up call shone a spotlight on U.S. deficiencies in the sciences and technology fields, and suddenly those areas got the benefit of attention by American policymakers and the public. Federal programs were developed, school curricula were adjusted, and funding programs were created to support the development in math, science, and engineering.

Many of the reforms that occurred during the post-Sputnik era are responsible for what is now the U.S.'s sport and physical activity/fitness culture, such as research into performance enhancement and the growth in government-funded programs aimed at providing physical activity options.

Critics of Title IX claim that its provisions have unfairly struck many men's sports, as athletics programs defund men's programs to come into compliance with women's opportunities. The unfortunate reality is that, despite the different ways schools can show compliance, some institutions have simply made it a financial game.

The sporting and fitness revolution of the 21st century

The 20th century saw significant changes and advancements in physical activity and sport. Yet who could have forecasted the way that physical activity and sport have become woven into society today? The choices for athletic involvement are staggering.

Community recreation leagues and travel leagues exist for almost any sport and activity that you can imagine. Group exercise classes of all types, both indoors and out, are common. Adventure experiences like hiking, rock climbing, mini-marathons, and triathlons are as available and common as are anything else out there. You name it, most likely a group is out there doing what you want. Just sign up and you can join.

In addition to so many more available choices for involvement in sport, opportunities are open for anyone, regardless of skill level or physical or emotional challenge. Much of the emphasis these days has been on the fitness benefits on participation rather than simply wins or losses.

Lifespan activities

Although master's level competitions for those age 35 to over 100 exist on the international scene, you don't need to compete to maintain involvement throughout your life. Lifespan activities offer participation, regardless of age and physical ability. These activities — like hiking, biking, swimming, and jogging — are ones that a person can carry out throughout his or her life.

Got a bum knee or back ache? You can still be active. If running hurts too much, go for regular walks or try the bike. If you can't jump for that rebound any more, take up racquet ball or swimming. Changing your activity to accommodate your body may be a way to ensure continued involvement. Plus wheelchair users, those who have cerebral palsy, and others with physical challenges can do the same; head to the next section for details.

Participation options for people with unique needs

Another field that increases inclusivity and allows those with more unique needs to participate is *adapted physical education.* People who lead these programs have knowledge of various conditions that have historically disallowed participation and fueled exclusivity to involvement. Examples of adapted physical activities include

- **Special recreational leagues:** These include wheelchair basketball leagues, for example, or *goalball,* a sport for the visually impaired.

- **The Paralympics:** The Paralympics are Olympic Games for those who have physical or intellectual impairments. People in the following categories are eligible to compete: amputees and those with limb deficiencies, those with cerebral palsy, those with intellectual impairment, and those who are visually impaired, just to name a few. Participation is ultimately determined by a committee that reviews each person's unique condition(s).

- **The Special Olympics:** The Special Olympics offers opportunities for those with intellectual disabilities.

Training for the trainers

As opportunities for athletic participation have expanded, so, too, has the need for qualified leadership. Today's coaches receive training in physiology, biomechanics, performance enhancement, sport psychology, and management, among other areas. Much of a coach's preparation has come out of the development of the academic fields that followed the boom of physical activity and sporting participation during the last century, when the popularity of athletic participation grew, and academic disciplines developed and became very focused.

Full-fledged fields of study like biomechanics, exercise physiology, and sport psychology are regular components of programs designed to help athletes of all skill levels enhance their performance. They have also allowed both athletes and the general population to significantly extend the longevity of their participation in sports and physical activity. The establishment of these scientific fields has led to an academic approach to sport performance enhancement and athlete participation. Stay tuned as you go through the rest of this book; you'll be exposed to kinesiology's core academic areas.

The current conundrum: A sports-obsessed nation with an unfit population

Americans continue to enjoy being spectators of sports, and at no time has our involvement in watching sport been greater. Although opportunities to watch sport have increased and resulted in staggering economic impacts, the increasing number of ways people can participate in sport and physical activity has been monumental. Not only have opportunities for children grown, but adult opportunities perhaps have grown even more.

Yet despite the opportunities to participate in both sporting and recreational activity, American is still a grossly unfit nation. Childhood and adult obesity are at all-time highs, and other *hypokinetic disorders* (diseases connected to the lack of physical activity) are costing the nation billions of dollars a year. Recognizing the poor health and its subsequent economic effects on the nation, the Healthy People initiatives of 2000, 2010, and 2020 seek to enhance the health of the nation by establishing goals for public health improvements.

What has become clear as a result of these initiatives is that many barriers to participation in physical activity programs are related to socioeconomic status and societal influences. For instance, not everyone can afford a gym membership, not all neighborhoods have green spaces to foster activity, and not all parks are safe places to play or work out. Additionally, many people just don't know where to start; for some, taking a walk just may not be safe, or paying for a club membership or class fee may be financially impossible. Work needs to continue to address socioeconomic factors that negatively affect participation, and helping the people who need assistance must happen at the ground level.

Still, something is out there for everyone. Our advice: It's less important what you do than just that you do something. Go for a walk around the halls of your office building or factory during lunch, take the stairs, park in the farthest parking spot . . . anything. Just move.

Part II
Exercise Physiology

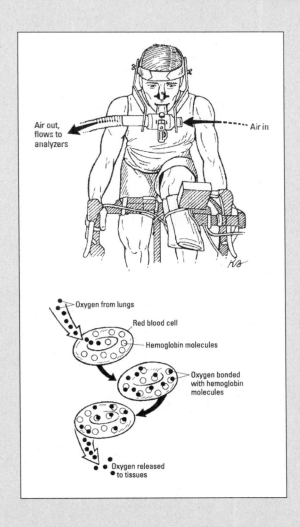

In this part...

✔ Discover the link between the mind and the body in human movement

✔ Uncover how body systems change during movement and how they adapt to continued movement and training

✔ Find out how muscles make movement possible, how you can train them, and how the complex systems of the body work together to help you perform your best

✔ Examine how the environment you move within affects your body during exercise and how you can adapt to your environment so that your performance isn't hurt

✔ Familiarize yourself with the way the body produces energy from fat, carbohydrates, protein, and even short-term, high-intensity energy sources

Chapter 3

The Brains behind the Brawn: Motor Control

Movement is a puzzle. Just walking across the room requires cooperation between muscles that move the legs, arms, and torso, plus balancing that noggin on top. Add to that changes in terrain, environment, and obstacles, and you can see how the brain has to coordinate many functions to execute movement. This is motor control.

Like any good relationship, communication occurs between the information coming to the brain from sensors that monitor what the body is experiencing (like pain, pressure, temperature, body position, and balance) and the information the brain is sending out in movement-related signals (like muscle contractions). These feedback systems help keep your body continually updated so that the movements you execute match the conditions you experience, and this constant communication is the basis of coordinated movement.

In this chapter, we explain how the brain and interconnected nervous system both sense body position and movement and initiate a movement response.

Introducing the Main Player: The Neuron

The primary functional part of the entire nervous system is the *neuron*. Neurons are small by themselves, but linked together they form the system by which signals are sent and received throughout the body. We each have billions of neurons in our bodies.

Neuron basics: Parts and functions

A neuron looks a bit like a tree with roots (see Figure 3-1). Familiarize yourself with the parts of a neuron and what they do:

- ✔ **Dendrites:** *Dendrites* are the thin, fingerlike sensory nerves that appear on one end of the neuron. They pick up signals from the *axon terminals* of other nearby nerves.

- ✔ **Cell body:** The *cell body,* the larger area containing the nucleus of the neuron, is responsible for keeping the neuron alive and functioning. The *nucleus* contains the genetic information for the neuron.

- ✔ **Axon:** The *axon* is the single nerve fiber that carries an electrical signal to the next neuron. Axons are bundled together, forming a nerve, and they can be very long or quite short. For faster conduction speeds, some axons are covered with a fatty coating, called a *myelin sheath.* The sheath allows nerve signals to jump along the outside of the nerve, hopping from open space to open space in the sheath (these spaces are called *Nodes of Ranvier*), much like a stone skipping across water.

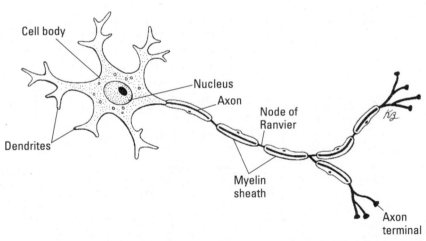

Figure 3-1:
The parts of
a neuron.

Cell body

Nucleus

Axon

Node of
Ranvier

Dendrites

Myelin
sheath

Axon
terminal

Illustration by Kathryn Born, MA

Neurons form connections to each other, creating branches of interconnections, which allow for information to pass across great distances and from various locations. So any particular neuron is actually getting inputs from a variety of neurons that are connected to it (at the dendrites). If the inputs are enough, they cause a stimulus in the neuron. If they aren't sufficient to cause a stimulus in the neuron, nothing happens. So one muscle may get a stimulus to contract, while a nearby muscle does nothing. It all depends on the sum of all the inputs.

Neurons are not all alike! Some have specific functions within the body. Some like to sense; others like to signal:

- **Sensory neurons:** Sensory neurons have dendrites that pick up sensory information from nearby neurons and send signals along their axons to the brain. This action is a bit like sitting in a crowded room, listening to conversations and then trying to explain to someone what you are hearing! In that scenario, you're the sensory neuron, picking up the signals (the bits you overhear) and passing them on.

- **Motor neurons:** As their name implies, motor neurons influence movement. Signals from the brain and spinal cord travel out to the muscles along motor neurons. These neurons influence how you throw the ball — or if you throw it at all. The signals can be excitatory (called *excitatory postsynaptic potentials*) or inhibitory (called *inhibitory postsynaptic potentials*).

- **Interneurons:** *Interneurons,* located in the brain or spinal cord, form a bridge between sensory and motor neurons. As we explain later, some movements can happen quickly after a sensation. These reflexes are hardwired into the body, using interneurons.

Neurotransmitters: The bridge over River Synapse

You may think that a sensory or motor neuron is like one long wire running through your body. Not so! In fact, neurons are more like small noodles, lined up in a row but not actually connected. Not connected? Nope. Which brings up the question, "How do signals get transmitted?" The answer: neurotransmitters!

Neurotransmitters are chemicals located in little sacs at the end of an axon terminal. When the nerve signal reaches these sacs, the sacs move to the edge of the axon and open up, releasing a chemical into the small space (called a *synapse*) between the two neurons. These chemicals float across the synapse and connect to the dendrites, stimulating the nerve to continue the signal. Think of the neurotransmitter as a ferry carrying a wagon across a river.

Different nerves have different neurotransmitters. Neurotransmitters that cause muscles to contract are not the same neurotransmitters that cause some nerves to either excite or inhibit. By having different types of neurotransmitters, the body can organize a variety of signals, rather than having one signal that tries to control everything.

Orders from Above: Motor Control

For movement to occur, the muscles need some coordinated instructions. Fortunately, we all have brains! The brain and associated nervous system provide the input to the muscles, stimulating them to contract and move.

In the following sections, we give you a tour of the central processing center (the brain) and the bundle of neurons (the spine) that transmit the signals the brain sends to the muscle.

The brain: The central processing center

As the central processing center, the brain receives sensory information (related to body position, balance, visual input, sound, and more) from all areas of the body and sends out information to the body in response. Some key areas of the brain include the following, labeled in boldface in Figures 3-2 and 3-3:

- **Cerebral cortex:** The wrinkled and wavy outer layer of the brain, this area has regions for both sensory information processing as well as motor function. Note the many different areas (limbic, primary visual, secondary visual, posterior parietal, auditory, and so on) within the cortex in Figure 3-2.

- **Sensory cortex:** Located toward the back (posterior) of the brain, the sensory cortex contains regions of sensory information like vision, taste, pain, temperature, and pressure.

- **Motor cortex:** The motor cortex is located toward the middle of the brain in the frontal lobe. It's a key area for initiating movement; it also helps coordinate movement (which is nice when you're trying to play darts and not put a hole in the wall).

- **Brainstem:** The brainstem is connected to the spinal cord and located under the cerebrum. It is partially made up of the pons and the medulla oblongata. This area has significant control over motor movements.

- **Cerebellum:** Attached to the brainstem, the cerebellum helps to smooth out muscle movement so that your jump shot is worthy of impressing your friends. Try threading a needle without the cerebellum making your movements smooth. Ouch!

- **Basal ganglia:** This area of the brain, shown in Figure 3-3, is important in learning motor tasks (like throwing a ball, swinging a bat, or even moving your eyes). This area can help smooth out movements and adjust movements as needed (as you do when you're making a jump shot and you have to adjust because someone tries to block you).

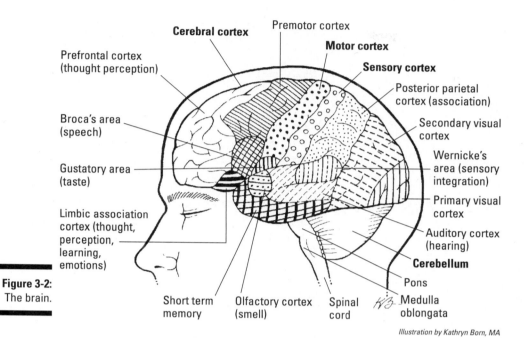

Cerebral cortex

Premotor cortex

Motor cortex

Sensory cortex

Prefrontal cortex
(thought perception)

Posterior parietal
cortex (association)

Broca's area
(speech)

Secondary visual
cortex

Gustatory area
(taste)

Wernicke's
area (sensory
integration)

Limbic association
cortex (thought,
perception,
learning,
emotions)

Primary visual
cortex

Auditory cortex
(hearing)

Cerebellum

Pons

Figure 3-2:
The brain.

Short term
memory

Olfactory cortex
(smell)

Spinal
cord

Medulla
oblongata

Illustration by Kathryn Born, MA

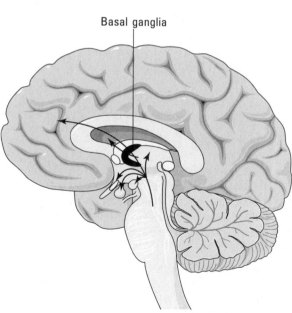

Basal ganglia

Figure 3-3:
The basal
ganglia.

Illustration by Wiley, Composition Services Graphics

Parkinson's disease: When neurons lose their steering

Although the nerves themselves may function just fine, the neurotransmitters that carry signals can sometimes be disrupted. Such is the case in Parkinson's disease. In this disease, the neurotransmitter dopamine begins to decline for reasons that are currently unknown. The primary location of the disease is in the basal ganglia, the area of the brain that helps control coordinated movement. As this neurotransmitter declines, signals to muscles as well as to other parts of the brain begin to falter. Movements become uncoordinated, and muscle tremors and rigidity are present. A cure has yet to be discovered, and most current treatments are designed to boost levels of the neurotransmitter and reduce the rate of decline.

The spinal cord: The autobahn of nerve signals

The spinal cord is a grouping of neurons that run the length of the spine. These neurons carry signals up to the brain and also carry signals from the brain to the muscles. The sensory and motor neurons are organized into horns on the spinal cord. The horns on the backside (dorsal) of the spine contain mostly sensory neurons, while those on the front side (ventral) of the spine contain primarily motor neurons. Interneurons, located on the ventral side of the spine, assist motor activity.

Feedback loops: Communicating between body and brain

As you move through the world walking, climbing, rolling, jumping, and doing whatever else sounds like fun, a continuous conversation takes place between what your body senses both externally and internally and the signals that the brain sends out. The brain interprets sensations and manages your motor efforts to match your desired movement. The communication between sensation and motor effort is referred to as a *feedback loop*.

Feedback loops function something like a thermostat. If you set the thermostat to 70°F and the room gets cool, the drop in temperature turns on the furnace, and the room heats up. When it hits 70°F, the furnace turns off. In a similar fashion, if you estimate the height of a step you're about to step on through visual clues, you prepare a motor response that lifts your leg and propels you upward to step smoothly onto the step. If you miss the step, sensory information about body position goes to the brain, and your next attempt is corrected.

Because the environment is continually changing and you're always moving, feedback loops are always in action, keeping you on track and walking upright!

The open-loop system

An open-loop system is one in which the activity is initiated, but nothing can be done to check or change the outcome. For example, you may program your sprinklers to come on every day at noon. But what if it rains? Your sprinklers still come on. An open-loop system works the same way.

The problem with an open-loop system is that no feedback exists to allow for control of the system; the system has been preprogrammed. The human body needs more control than that. For that reason, most of the feedback systems in your body are closed-loop systems.

The closed-loop system

A closed-loop system is more like a thermostat, in that sensory feedback helps guide the brain's response. If your body temperature goes up, the hypothalamus in your brain triggers sweating to cool you off. If your blood sugar goes up, receptors sensitive to this rise provide the brain information that provokes a stimulus to the pancreas to release insulin to help lower blood sugar.

Examples of closed-loop systems include the following reflexes, which you can read more about in the later section "Hardwiring the nervous system: Reflex control" and in Chapter 10:

- ✔ **Muscle spindles:** Stretch the muscle too fast, and the feedback stimulus causes the muscle to contract against the stretch.
- ✔ **The Golgi tendon:** Put the tendon under too much tension, and the feedback initiates an inhibition of muscle contraction to help ease the tension.

Your Place in Space: Sensory Information and Control of Movement

To have an effective feedback system, you need both sensory information coming into the brain and accurate motor signals being sent out to the muscles. Sensory information comes from a variety of systems, some of which provide an immediate motor response (a reflex), while others add to a list of inputs that help you coordinate some pretty sophisticated movements!

Where did I put my hand? Sensing body position

Try walking in a dark room sometime. What is your first instinct as you try to get a sense of your movement? You reach out to touch something, you search for any visible light to get a sense of up or down, and you feel the incline of the floor. Why? Because movement and performance of activity are closely linked to the sensory tools that constantly tell you where you are in space. Put together, these senses — touch, vision, and proprioception — help you perform activities with precision.

Touch

Skin may look like it's just tissue and blood vessels, but it's filled with sensors that provide vital information:

- **Touch:** Light-touch sensors (called *Meissner's corpuscles*) provide a tactile sense that helps you apply the appropriate amount of pressure (like holding a baseball with just enough grip to throw a knuckleball). They are sensitive to very light touch and vibration.

  Heavier touch can be sensed by sensors, called *Pacinian corpuscles,* that are deeper within the skin. These sensors help distinguish the difference between rough and smooth.

- **Temperature:** Temperature sensors indicate hot and cold, information that can help you determine whether you want to take a sip of your coffee or blow on it awhile!

- **Pressure:** Pressure sensors on the feet tell you whether you are standing and how balanced you are on your feet, for example. They also help provide feedback as you change positions (like making a cross dribble fake in basketball!)

When you're swinging a bat, a golf club, or a tennis racquet, the senses of touch and pressure as you hold and hit are essential for performing well.

Vision

Vision sensation is an essential part of coordinated movement. You can always spot new dancers, for example. They're the ones staring down at their feet rather than looking at their partners. You'd almost think they need to see each step to be sure they're moving!

Just how useful is visual feedback? It depends on what you're looking at. Although the vision sense may be similar for individuals, how people use their visual sense may not be. Consider these examples:

- **Boxing:** An untrained boxer may watch the opponent's gloves, anticipating a punch to avoid. A trained boxer, on the other hand, may watch the opponent's trunk, hips, and shoulders in order to anticipate a punch. In the latter case, the boxer may easily avoid the punch; in the former, he may be eating a glove sandwich!

- **Baseball:** A novice baseball or softball player may watch for the ball as it approaches the plate. Expert players see the release point from the arm of the pitcher, which provides information about the angle and even the type of pitch that may be coming.

Competitors often try to disguise or deceive by providing misleading visual feedback to their opponents. They do this because they know that visual senses are essential components for guiding motor responses.

Proprioception

Have you ever been asked to close your eyes and touch your nose? Hopefully it was just for fun! How did you know where your nose was? Or how did you know where your arm and finger were? Even without visual cues, you have internal sensors, *proprioceptors,* that provide position sense about your limbs, head, and trunk. Like an internal GPS, proprioceptors track position, movement, and velocity.

A few key proprioceptors are the *muscle spindles,* which sense change in length and velocity of the muscle fibers; and the *Golgi tendons,* which sense the amount of tension in a muscle. We explain these proprioceptors in the next section.

Which way is up?

The process of standing isn't as easy as it sounds. Sensory cues from pressure receptors on the feet tell you where your center of balance is. Your eyes help you orientate to the upright position, and your inner ear provides more feedback regarding your position.

But what about an astronaut? In a weightless environment, every way is up! No sensory stimulus exists to provide body position or center of balance. This lack of information makes adjustment to space a difficult thing, often leading to nausea for days. Over time, however, the astronauts do adjust.

Yet when they come back to Earth, the sensory receptors have lost their "touch," so to speak, and don't initially provide enough sensory feedback to help orient the astronauts, leading to significant balance issues. Within a few days, these senses return, allowing the astronauts to once again stand up and stare at the stars.

Hardwiring the nervous system: Reflex control

If a building catches on fire, it's nice to know that the sprinklers will come on automatically. No one needs to be called for a decision! Likewise, if your tendon was being pulled on so hard it could tear from the bone, it would be nice if your body automatically stopped the muscle from pulling harder without waiting for a decision from the brain. Such is the role of a *reflex,* a preprogrammed sensory-motor control system that is designed to protect the body. Read on to find out about a couple of key reflexes.

Muscle spindles

When you think of spindles, think of speed! Muscle spindles are sensory fibers intertwined with muscle fibers. They look a bit like a slinky wrapped within the muscle and moving with every change in muscle length.

Spindles sense the change in length and velocity of the muscle fibers. Because muscle spindles are sensitive to changes in muscle length and speed, what happens when you rapidly stretch a muscle? You trigger a stretch reflex. The spindles sense the rapid stretch of the muscle, and sensory nerves send this signal to the interneurons of the spinal cord. A stimulus to the muscle is initiated, causing muscle contraction. At the same time, a signal is sent to the opposing muscle to relax so that movement can take place.

You can think of the stretch reflex this way:

rapid stretch = reflex muscle contraction

When the doctor taps your knee at the tendon, she is trying to induce a small, rapid stretch of the quadriceps muscle. The rapid stretch (like plucking a guitar string) results in a reflex contraction, and you notice a slight kick of the leg.

Golgi tendon organ

When you think of the Golgi tendon, think tension. The Golgi tendon monitors the amount of tension in the tendon that attaches the muscle to the bone (much like a spider feels tension in its web while waiting for the next meal to fly in).

How can tension occur at the tendon? In two ways:

- **Lifting a heavy weight:** The force required by the muscle to lift a heavy weight pulls hard on the tendon, creating tension.

- **Stretching a muscle:** When the muscle is placed under a prolonged stretch, the tendon experiences tension.

Plyometrics: Training the muscle spindle for peak performance

One way to induce the muscle spindle to fire in the leg and gluteal muscles is by jumping down from a box. The landing motion creates a rapid stretch in the muscles, causing the muscle spindles to fire. When the person then tries to jump immediately after the landing, he or she gets a "boost" from the additional muscle firing due to the spindles.

Additionally, with repeated training, you can actually speed up the time between the spindles sensing the stretch and the muscle firing. This enhancement can make a big different in athletic performance when tenths of a second count!

A reflex response occurs in response to tension. Sensory nerves attached to the tendon sense tension and respond by inhibiting the muscles from contracting, causing the muscle to relax.

Try doing a hamstring stretch and hold the stretch position so that the muscle is placed under constant tension. Within 20 seconds, you'll notice that the muscle starts to relax! You have discovered the Golgi tendon reflex!

Now suppose that you're lifting a series of weights, each one heavier than the last. As you get to your heaviest weight, the amount of tension on the muscle and tendon is so much that the Golgi tendon is activated. As a result, your muscle can't exert any additional force (even if the muscle had the ability) because the Golgi prevents additional muscle recruitment.

If you have ever driven (or followed behind) a gas-powered golf cart, you know these vehicles go pretty slowly because a device prevents them from going faster. If you removed that device? Zoom! Well, the Golgi tendon acts in a similar way, preventing the muscle from creating too much tension. However, you can modify how the Golgi tendon responds. With training, you can increase the level of tension needed to activate the Golgi tendon, thus enabling the muscle to exert more force and making you stronger! You don't even need to grow more muscle; you're just using more of what you have.

Using muscle spindle and Golgi tendon reflexes to enhance performance

Your built-in reflexes are meant to protect you. And if you trigger them correctly, you can use them to enhance performance. However, in some instances and with some movements, these reflexes can hurt performance. Here's what you need to know:

✔ **Should you hold a stretch or bounce a stretch?** Hold stretches and bouncing stretches activate different reflexes. During the hold stretch, the Golgi tendon is activated and the muscle relaxes. Yay! Nice stretch.

During the bouncing stretch, the muscle spindles are activated (due to the rapid stretch). The reflex is a muscle contraction! Ever try to stretch a muscle that is contracting? Ouch! That can damage the muscle. Not the best stretching idea.

✔ **Should you do some mini-jumps to warm up or should you stretch?** Mini-jumps activate the muscle spindles, which respond by activating muscle contraction. Yay! More muscle to jump with! Nice exercise. However, if you hold a stretch of those muscle just before the activity, the Golgi tendon responds by inhibiting muscle firing. Looks like you won't be jumping too high this time.

Threading the Needle or Shooting a Free Throw: Coordinating Movement

If you have ever played a video game that involves fast driving or flying, you know that when the path gets too complicated you have to slow down to navigate. Things just come too quickly to make steering adjustments fast enough! Likewise, many human activities, like hitting a ball or putting a key into a keyhole require speed, accuracy, or a combination of the two. In the next sections, we explain how these are coordinated.

Making the speed-accuracy trade-off

Most sport activity requires targeted movements at a particular speed (or as fast as you are able to move); pitching a ball to a batter, swinging a golf club to hit a ball, and throwing a football to a running receiver are examples. If a skill is particularly difficult because of the high level of accuracy required, like swinging a golf club to hit a ball, you sacrifice speed (slow down) in order to gain accuracy. This is known as the *speed-accuracy trade-off*.

In this trade-off, anytime you increase the speed of a movement, you reduce the movement's accuracy. Likewise if you increase the accuracy needed for a movement, you sacrifice the speed at which you can perform.

So how can a professional golfer swing a club with such a high velocity and still hit that tiny white ball? Training! Through repetitive training, you can condition your motor system to create accurate movements at very high speeds.

Following the phases in a movement

Performing an activity is a sequence of a number of events. The first phase is preparation, followed by the initial movement, followed by a terminal stage of movement. Speed and accuracy (covered in the preceding section) are dependent upon your sensory motor system making adjustments throughout all the phases, from beginning to end:

- **The preparation phase:** In the preparation phase, you have decided upon the movement you want to initiate. If you were hitting a golf ball with a club, for example, your visual senses determine the distance to the ball and the position of the ball on the tee. Pressure sensors in your feet determine your balance on the ground, the degree of traction you have to use for force, and the direction of the intended flight path of the ball.

  As you begin to draw the club back (your backswing), proprioceptors indicate where your arms are in relation to your body, the position of the club, and the velocity of the backswing, while your eyes focus on the upcoming impact point on the ball. Lower limb pressure and position sensors provide information about your windup phase. At this point, your muscles are stretching and your body is placed in a position where it can unwind in a forceful fashion.

- **The initiation phase:** In the initiation phase, the speed of movement comes into play. Your vision maintains contact with the target object while your motor control system initiates an explosive movement to swing the club. Little adjustment in the swing is possible during this movement because it is done at such a high velocity. Just swing, baby!

- **The terminal phase:** The terminal phase occurs just before the ball is hit and the point of contact. Using visual senses, your central nervous system can allow for subtle changes in the swing at the point of contact to adjust trajectory.

So think of all the minor swing adjustments that occur during the preparation, initiation, and terminal phases of a golf swing. Improvement in both speed and accuracy can happen only over time and with repeated swing practice. Of course, golf uses different clubs, so the swing training needs minor alterations for each club and each ground condition. No wonder golf is such a hard sport!

The field of biomechanics studies the physics of body movements (as well as the club you might be swinging) and examines the phases of movements and the aspects of the body in motion. Chapter 11 contains information on analyzing the biomechanics of movement.

Coordinating two arms: Bimanual coordination

You body likes to move its limbs (arms and legs) in a coordinated fashion (called *symmetrical coordination*), in which both arms and both legs do essentially the same thing. But what happens, for example, if the arms have to move in different ways (called *asymmetrical coordination*)? Playing the violin can be quite complicated when one hand is forming cords while the other hand (and arm) is moving a bow across the strings!

These types of movements are difficult simply because limbs prefer to work in unison. They like symmetrical movement! Consider these points:

- ✔ The more complex a movement is for a limb, the slower the limb will move.

- ✔ The limb doing the more complex movement influences the movement of the limb doing the less complex movement.

- ✔ With a new complex movement that requires each limb to move separately, both limbs initially want to do the same thing.

Even though limbs want to work in unison, you can learn to move each limb separately. Here's how:

- ✔ **Practice.** Start with slow movements and work up to the speed required. Initially, the slower movements allow for the proper sequencing of the movement. You can add speed later.

- ✔ **Get feedback.** Find a coach who can provide instruction on proper position and initiation of movement. For the novice, attaining the proper form and position is difficult, but the practiced eye of an expert can help correct movements before they become learned in the wrong way.

- ✔ **Try breaking the movement down into its components and practice each component separately.** Initially, practicing the individual components of a movement can help you develop the skill. After the motor skills for each part of the movement have become ingrained, you can focus on sequencing the parts together into a more complex movement. For example, you may want to first practice tossing the volleyball in the air a bit, then practice stepping and hitting the ball, and then finally put these actions together to master your serve.

Now you can see why walking is easy (or it *seems* easy), but dancing can be a challenge. One movement (walking) has been learned, while the other (dancing) may involve first stepping forward, then to the side, then forward again. Both movements are similar, but the movements associated with dancing haven't been ingrained. No wonder new dancers often feel like they have two left feet!

Come on, baby, do the locomotion: The rhythm of walking

Upright walking is unique to humans, and most of us get the basics down by the time we're 1 year old. Locomotion is programmed into our basic nervous system functions (we don't need to think about it). Sensory signals then help shape or modify our movements, specific to the demands. When talking about walking locomotion, the term frequently used is *gait*.

Watch how people walk. Do they simply stride with arms at their sides in a stiff upright position? No! Instead, you'll notice a rhythm and relationship between the leg stride, arm swing, and trunk rotation. Walking is like dancing, as we explain in the next sections.

 Most research indicates that the coordination of walking is a centralized (nervous system) control, where a pattern of signals is generated for normal walking. For this reason, a disruption in walking patterns could indicate a problem with the central nervous system.

Leg movement

Leg movement occurs in two primary phases, each completed separately by the legs (see Figure 3-4):

- **Stance phase:** During this phase, the limb is extended and placed in contact with the ground. During ground contact, the push to move forward is generated.

- **Swing phase:** Following the stance phase is the swing phase, during which the limb is flexed to leave the ground and then brought forward to begin the next stance phase.

Out of phase: Parkinson's disease and locomotion

Normal gait consists of upright posture, steady stride between the stance and swing phases, and arms rhythmically (and relaxed) moving in opposition to the stride. Due to the neurological disruption caused by Parkinson's disease, the gait is disrupted, and normal movement is no longer evident. In people with Parkinson's disease, you see the following:

- A slowness of movement

- Slow initiation of walking and stooped body position

- Arms rigid and at the person's side while running, due to muscle rigidity

- Shuffling steps rather than the classic stance-to-swing transition

All these symptoms are indicators of the disruption in central nervous system function. Medications can help alleviate some of the symptoms.

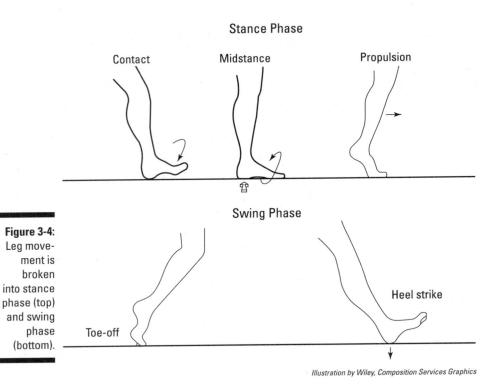

Figure 3-4: Leg movement is broken into stance phase (top) and swing phase (bottom).

Each leg is coordinated in the two phases separately, and faster movement is completed generally by shortening the stance phase.

Arm movement

The arms move in a synchronized fashion in relation to the legs to provide balance and propulsion. For slow walking, the arms swing twice for every leg stride, but at faster speeds, the arm-to-leg ratio is 1:1 (one arm swing to one leg swing).

Trunk movement

The trunk serves to both stabilize the upper body during the leg propulsion phase (the latter part of the stance phase) and also to position the pelvis in a better position to maintain balance while executing arm and leg movements.

Chapter 4

Keeping the Big Wheel Turning: Exercise Metabolism

. . . .

In This Chapter

▶ Understanding the systems by which your body produces energy (ATP)

▶ Measuring and using oxygen consumption: VO_2

▶ Training your metabolism

. . . .

*Y*our body runs on one fuel source: ATP. But because you can store only a small amount of ATP (like a battery with little charge), as soon as you begin to use it up, you need to create more. Fortunately, you can do so by using your body's ATP generators!

Some of these systems kick in fast and furious, providing an almost instant supply of ATP; others provide energy at a slower rate. Some get depleted quickly; others can go forever. Some sport activities and movements use one system more than the others. In this chapter, we explain how your body's ATP systems work and note which activities use one system more than another.

Introducing The ATP-PC Energy System: Give Me Energy Now!

Your life depends on the energy you get when you break chemical bonds, specifically the bonds holding together a molecule called *adenosine triphosphate,* or *ATP.* Adenosine is connected to three phosphates by high-energy bonds. These bonds hold the energy that drives all the biological actions in your body. To produce energy, you just need to get the bonds to break!

Breaking (chemical) bonds

An essential component that drives chemical reactions is a *catalyst*. In the human body, these catalysts are *enzymes*. It takes a reaction with water and the help of a special enzyme — ATPase — to liberate energy from ATP. Once the energy is liberated (along with a phosphate being lost), you end up with *adenosine diphosphate,* or *ADP* (the *di* part of *diphosphate* means "two").

When ADP is present, systems begin to turn on to replenish ATP. Here's what's happening:

$$ATP + H_2O \rightarrow ADP + P_i = ENERGY$$

Where is ATP stored? Well, if it's used for movement, where do you think you'd put it? In the muscle! But strangely, muscle stores only enough ATP for about two seconds of work. That's like having $5 in the bank — certainly not enough to pay the bills. Fortunately, your body has a way to produce the ongoing energy you need.

Replenishing energy as you use energy: The air compressor analogy

To understand how the human body's energy system works, think of an air compressor. An air compressor is a big tank that holds air that you can use to fill tires or run machinery. Attached to this tank is a small motor that compresses air and a gauge that shows the amount of air pressure in the tank.

Now imagine that the tank is full of compressed air and the motor is turned off. How can you make the compressor turn on? Simple: You use some air! When you use air, the pressure in the tank drops, and that drop in pressure signals the motor to turn on to compress more air.

Your body's energy system follows essentially the same principle. You have only two seconds of ATP stored in your body, but when you use it, you turn on your body's energy systems to start making ATP!

Three primary systems provide ATP:

- ✓ **Phosphocreatine (ATP-PC):** This system provides an immediate boost of energy that lasts only a few seconds.

- ✓ **Anaerobic glycolysis:** This systems provides energy for activities lasting longer than a few seconds (closer to five minutes) but that still need a lot of ATP quickly.

- ✓ **Oxidative (aerobic) system:** This system provides energy for activities that don't need ATP at very high rates but need the ATP to last for a very long time without fatigue (like your entire life!).

As you can see, the systems that make ATP provide energy at different rates. The moment you start to move, you use your small stores of ATP and turn on *all* the systems that make ATP. Because your metabolism always needs ATP, you are always making it and always using it! As a result, you can get energy for long walks or high jumps without having to shift gears. The energy will be there for you! The following sections explain these three energy systems in detail.

Phosphocreatine: An Immediate Source of ATP

If the high-energy bond of ATP is broken for activity, that energy needs to be replaced if you want to continue moving. One way your body replaces the used energy *fast* is to "steal" it from another substance. *Phosphocreatine* is the most immediate source of energy to remake ATP (see Figure 4-1). Phosphocreatine is stored within your muscle, and you have enough for about ten seconds of all out, high-intensity exercise.

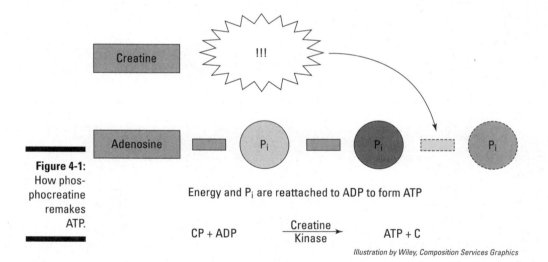

Figure 4-1: How phosphocreatine remakes ATP.

Energy and P_i are reattached to ADP to form ATP

$$CP + ADP \xrightarrow[\text{Kinase}]{\text{Creatine}} ATP + C$$

Illustration by Wiley, Composition Services Graphics

As noted previously, phosphocreatine produces enough energy for about ten seconds of your hardest effort. Say you're running a 100-meter race. The race begins, and you take off. You're breaking down ATP fast. But as fast as you use it, phosphocreatine provides the energy to remake ATP. Essentially, it's your friend handing you $100 bills as quickly as you spend them. Yet as you run along, you continue to deplete your phosphocreatine stores. When it runs out, ATP supply drops quickly, as Figure 4-2 shows, and you must slow down.

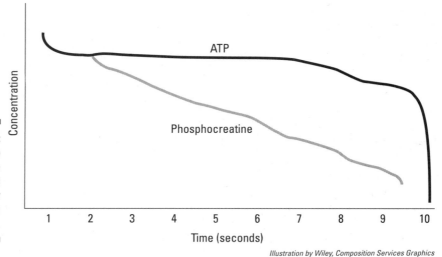

Concentration

ATP

Phosphocreatine

1   2   3   4   5   6   7   8   9   10

Time (seconds)

Figure 4-2:
Phospho
creatine
stores
last only
about ten
seconds.

To better understand how the phosphocreatine system works, imagine that you have $100 in your pocket and a friend who replenishes your money as you spend it. You spend $10 on lunch, for example, and immediately, your friend hands you $10. You spend another $50 on some clothes. Immediately, your friend gives you $50! Wow! As long as your friend doesn't go broke, you can keep spending! But when the friend goes broke, you're both out of luck.

Creatine loading: Effective aid or expensive urine?

Because phosphocreatine is so important for high-intensity exercise, athletes want to store as much of it as possible. The human body can create some of the components, like creatine, from amino acids. However, some research has shown that supplementing the diet with creatine may also provide positive effects. In laboratory testing, creatine supplementation showed positive improvements in high-intensity exercise performance (producing a better effort at weight training, for example, or leading to increased strength and longer efforts at sprint cycling, resulting in improved sprint time).

The benefits seem to vary: Some subjects responded to the supplementation; others didn't. Side effects appear to be minimal (nausea and possibly cramping). In addition, the addition of creatine to the cells pulls water with it, causing some water weight gain — an effect that may be beneficial for those needing to cool themselves with sweat but deleterious to those who need speed (the added weight can slow them down). So the jury is still out on the effectiveness of creatine supplementation.

High-intensity activities that last under ten seconds — activities like the sprinting for 100 meters, high jumping, running to first base in baseball, swinging a bat or golf club, or lifting a very heavy weight (one that you can lift maybe only five times) — all get their ATP mostly from phosphocreatine.

Anaerobic Glycolysis: Fast Energy with a Price

Wouldn't it be nice to be able to generate enough ATP to run hard for longer than the ten-second burst of phosphocreatine? Well, you do have a system of ATP production that works very quickly to give you ATP. Using the sugar in your body, you can produce a lot of ATP pretty fast (not as fast as the phosphocreatine system can produce ATP, but still plenty fast). However, this ATP production method has a side effect of sorts: It leads to fatigue.

ATP supplies are used up within two seconds of exercise. This loss of ATP immediately turns on the system within the cell that can use the simple sugar, glucose, to provide energy to form more ATP. The chemical reactions that accomplish this take place within the cell.

Your starter fuel: Glucose and glycogen

Because humans run on ATP, you'd think that we should just eat ATP. Not possible! Instead, the energy for ATP is tied up in other molecules. One is the carbohydrate glucose. Your body has the necessary enzymes to rearrange a molecule of glucose so that the energy within the molecular bonds can be used to remake ATP. How sweet is that?

Many versions of sugars exist, but the simplest is *glucose,* the primary component of starches, or carbohydrates, like pasta, grains, rice, and sugars, for example. Glucose is a 6-carbon molecule. Plants create glucose through photosynthesis. We humans use glucose for energy.

Although you carry a small amount of glucose in your blood, you actually store glucose in the muscle and liver in the form of *glycogen.* Think of glycogen as a glucose snowball — a multitude of glucose units connected together. Your body has about 2,000 calories of glycogen — enough to run about 20 miles.

Getting glucose into the cell

Because glycogen is kept in two locations (the muscle and liver), the process of getting glucose ready for ATP production varies, depending on the location of the glycogen:

✔ **From the liver:** Glycogen is broken apart into individual glucose units (a term call *glycogenolysis*). This glucose is dumped into the blood so that it can be transported to the muscle or to other cells that need it. (This is why we have blood glucose.) From there, the glucose gets into the cell by way of special proteins that act like gateways (think of the revolving doors at a fancy hotel).

ATP is required for glucose to get into cells, so you actually use energy to start the process of creating ATP. The end result is a glucose molecule with a phosphate attached (called *glucose 6-phosphate*) — just what you need to start making some ATP quickly inside the cell.

✔ **From the muscle:** Location is everything! Because some glycogen already exists in the muscle, you simply need an enzyme to break the glycogen snowball apart and grab a floating phosphate. The result? A glucose 6-phosphate that didn't cost you any ATP.

Glycogen supercompensation: Carb loading for performance

Having as much glycogen as possible stored in the liver and muscle before any big activity or event (a marathon, the Tour de France, and so on) is clearly advantageous. Normally, you store glycogen after a workout, using a glycogen-making enzyme called *glycogen synthase*, along with carbohydrates in the diet. Briefly storing more glycogen than normal is possible, however. The recipe for *loading*, or *supercompensating*, is very simple:

1. **Reduce your training volume and intensity.**

   Known as a *taper*, this action reduces your use of glycogen. Filling the gas tank is easier if you aren't using as much gas, right?

2. **Increase your dietary intake of carbohydrates.**

   Maybe bump it up to 60 percent to 70 percent of your diet. The enzymes for glycogen synthesis store the extra carbs as glycogen.

3. **A day or two before the big event, reduce your training even further and increase your carbs to at least 70 percent.**

   Doing so drives maximal glycogen supercompensation.

The result? You have extra stores of glycogen, you feel rested, and you probably can run faster and longer than you ever did during training.

Cooking up ATP, oxygen free: Anaerobic glycolysis

The cell is a bit of a soup, full of nutrients and enzymes. Chemical reactions can take place quickly in this environment and help produce ATP at a very fast rate. Because oxygen isn't used, the process of breaking down glucose is *anaerobic* (which means "without oxygen").

Two primary steps are involved with anaerobic glycolysis. The first step requires an investment of energy, but the second step doubles your energy. If you want to make money, you have to spend a little, and that's what happens here (see Figure 4-3):

- ✔ **Step 1: Invest energy and prime the pump:** In this first phase, glucose is taken into the cell and "trapped," by attaching a phosphate and converting glucose into glucose 6-phosphate. Capturing and trapping the glucose takes energy (ATP) and an enzyme. Later, another phosphate is added and another ATP used. This process gets the molecules in a position to produce ATP by converting it to fructose 6-phosphate.

- ✔ **Step 2: Produce double the energy:** In this phase, the molecule splits in two, and each one goes through a series of reactions, starting from glucose 3-phosphate (G3P), to 1–3 biphosphoglycerate (1–3 BPG), to 3 phosphoglycerate (3-PG), during which you generate two ATP). After water removal, phosphoenolpyruvate (PEP) is converted to a strong acid, *pyruvic acid,* and two more ATP are generated. You just doubled your money — you went from 2 ATP to 4 ATP — and have a net gain of energy!

The reason you can get one more ATP is because muscle glycogen doesn't take energy to trap the glucose, so one less ATP is invested!

This outcome almost seems too good to be true: fast energy, double your energy . . . so what's the catch? The end result of glycolysis, *pyruvic acid,* or *pyruvate,* is a strong acid that very quickly causes fatigue.

One way to slow the acid build up is to convert the pyruvate to a less-acidic acid — *lactic acid.* An enzyme (lactate dehydrogenase, or LDH) converts pyruvic acid into lactic acid:

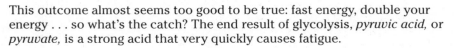

Pyruvic acid ——— Enzyme (LDH) ——— Lactic acid

Unfortunately, lactic acid can be just as bad, as we explain in the next section.

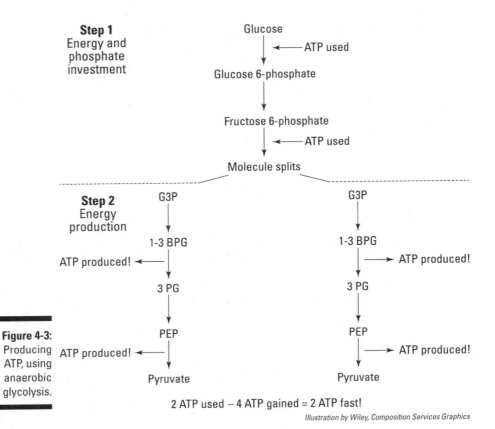

Step 1
Energy and
phosphate
investment

Glucose

← ATP used

Glucose 6-phosphate

Fructose 6-phosphate

← ATP used

Molecule splits

Step 2
Energy
production

G3P

1-3 BPG

ATP produced! ←

3 PG

PEP

Figure 4-3:
Producing
ATP, using
anaerobic
glycolysis.

ATP produced! ←

Pyruvate

G3P

1-3 BPG

→ ATP produced!

3 PG

PEP

→ ATP produced!

Pyruvate

2 ATP used − 4 ATP gained = 2 ATP fast!

Illustration by Wiley, Composition Services Graphics

The metabolic bad boy: Lactic acid and fatigue

The unfortunate side effect of anaerobic glycolysis is the accumulation of lactic acid. Your body doesn't respond well to acid. Proteins begin to break down, and cells can die. To prevent too much acid from building up, a number of fatigue-related changes take place when lactic acid accumulates from exercise. These changes are designed to slow you down and prevent damage.

When we talk about acid in this context, we're really talking about a particular ion. Think of the unit of acidity, pH. pH stands for the "power of hydrogen." The hydrogen ion (H^+) is what makes acid, well, acid. The concept can be confusing sometimes, because a higher concentration of H^+, or more acidity, is reflected by a lower pH value. So the accumulation of lactic acid really means the accumulation of hydrogen ions and a drop in pH.

Understanding how H⁺ causes fatigue

Following is a step-by-step description of how H^+ can cause fatigue; Figure 4-4 shows this process:

1. **Hydrogen ions block muscle contraction.**

   H^+ ions compete with another ion, calcium (Ca^{++}) for the affection of a protein (troponin). Calcium normally binds to troponin in the muscle to cause muscle contraction. As H^+ builds up, it competes with Ca^{++} for troponin binding and starts to block contraction. The result? Less force produced and fatigue.

2. **Hydrogen ions slow nerve signals to the muscle.**

   Normally, nerve signals can skip along a nerve like a stone across water. H^+ in the cell slows the skipping down. As a result, you can't get a coordinated signal to the muscles. Uncoordinated signals interfere with motor skills. Your running strides are altered, and you experience more fatigue.

3. **Hydrogen ions block an enzyme necessary for anaerobic glycolysis.**

   Early in the anaerobic glycolysis process of making ATP, a particular enzyme, *phosphofructokinase,* is very sensitive to H^+. When H^+ levels rise, phosphofructokinase slows its ability to help run glycolysis! When this happens, you can't make more lactic acid. Of course, you can't make more ATP either. Result? Even more fatigue.

4. **Hydrogen ions cause pain.**

   You may have felt the pain of lactic acid in your muscles when you work hard. And when the pain is high, you reduce your effort (which means a reduced motor stimulus leaving your brain); that is also fatigue.

Lactic acid causes fatigue at the brain (called *central fatigue),* as well as at various locations around the muscle and cell *(peripheral fatigue).* Head to Chapter 3 for a more in-depth discussion of the brain's role in movement.

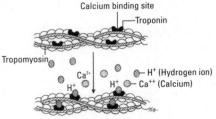

Source of fatigue # 1

Lactic acid interferes with muscle contraction

Calcium binding site

Troponin

Tropomyosin

Ca^{2+}

H^+ (Hydrogen ion)

H^+

Ca^{++} (Calcium)

H^+

Hydrogen ions from lactic acid compete with calcium for binding sites. This blocks muscle contractions.

Source of fatigue # 2

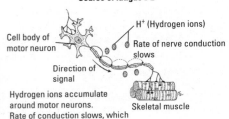

H^+ (Hydrogen ions)

Cell body of motor neuron

Rate of nerve conduction slows

Direction of signal

Hydrogen ions accumulate around motor neurons. Rate of conduction slows, which slows signals to the muscle and results in muscle fatigue.

Skeletal muscle

Source of fatigue # 3

Depleting stores of glycogen

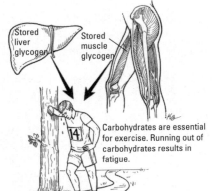

Stored liver glycogen

Stored muscle glycogen

Carbohydrates are essential for exercise. Running out of carbohydrates results in fatigue.

Figure 4-4: How lactic acid causes fatigue.

Illustration by Kathryn Born, MA

Recovery from exercise: Getting rid of lactic acid

Because lactic acid is such a major player in fatigue (and fatigue means poor performance), getting rid of lactic acid fast is desirable if you want to get back in action quickly. Here are some suggestions:

✔ **Breathe it off.** Thank goodness for chemistry! Blood contains a substance called *bicarbonate* (HCO_3^-). When you breathe, the HCO_3^- in your blood can combine with the nasty H^+ to form a weaker, and less nasty, acid called *carbonic acid*.

$$H^+ + HCO_3^- \longrightarrow H_2CO_3$$

Can you see two substances in there? Well, when blood passes through the lung, it's converted into carbon dioxide and water:

$$H_2CO_3 \longrightarrow H_2O + CO_2$$

The extra CO_2 makes you breathe harder. You've probably found yourself breathing hard when you've pushed the intensity a bit too much. Maybe it's because lactic acid is building up!

✔ **Use your muscles.** As you'll see when you study muscles, muscle fibers favor either fast or slow ATP production. Aerobic fibers, known as *slow twitch* muscle fibers, are unique. They actually use lactic acid just like glucose and produce ATP! Slow twitch fibers are used mostly during light exercise. So when the big sprint is over and all that lactic acid has built up, what should you do? Walk! Light jog! If you use your slow twitch fibers, you'll get rid of the lactic acid more quickly.

The Oxidative (Aerobic) System: It Just Keeps Going and Going

Mitochondria are nature's batteries. These organelles contain enzymes that can rearrange molecules through a number of steps to ultimately create ATP. The reason mitochondria are so good at producing ATP is because, as long as they have a source of fuel and oxygen available, the only waste they produce is carbon dioxide and water, which you can breathe out.

Mitochondria (see Figure 4-5 for an image of a mitochondrion) are located in the cell. From a standpoint of exercise, you'll find that they may be more numerous in some areas than in others. As Chapter 10 explains, different muscle fibers have different characteristics; some are aerobic muscle fibers and have a lot of mitochondria; others are anaerobic fibers and have few mitochondria.

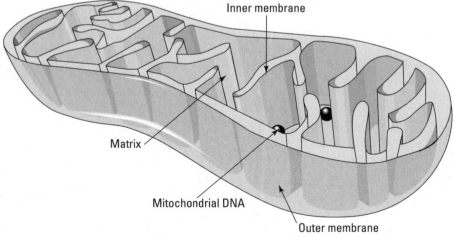

Figure 4-5:
A mitochon-drion.

Inner membrane

Matrix

Mitochondrial DNA

Outer membrane

Illustration by Wiley, Composition Services Graphics

In the mitochondria, chemical reactions occur that take fat and glucose and produce H⁺ ions. These ions create the power to make the ATP that feeds your muscles.

Mitochondria produce ATP much like a battery produces energy. To understand these chemical reactions, picture a battery. A battery has a positive and a negative charge at either end. The flow of electrons across this electrical gradient is what runs your radio. Inside the battery, positive charges are created, usually using an acid that produces hydrogen (H⁺) ions.

Aerobic metabolism: Making ATP with glucose, fat, and protein

You can make energy from the foods you eat. This process, called *aerobic metabolism,* involves a series of complex chemical reactions — too complex for the purposes of this book. However, you may find it useful to take a step-by-step look at the basics behind the creation of ATP from the nutrients in foods.

For glucose and fat to be used within the mitochondria, they are rearranged into various molecules that can "plug into" the aerobic metabolic pathway and act much like kindling does to feed a fire. The "fire" of ATP production in the mitochondria is a sequence of molecular rearrangements, and glucose, fat, and protein serve as the starting point.

B vitamins: The energy vitamins?

You may have heard that B vitamins are the "energy vitamins," a phrase that may lead you to think that, just by consuming these vitamins, you'll have energy. Unfortunately, that isn't the case. You get energy through the aerobic and anaerobic systems.

Still, two B vitamins do play a very important role in making ATP. As the section "The Krebs cycle: The body's hydrogen producer" explains, H⁺ ions produced by the Krebs cycle get carried, or shuttled, to the electron transport system. Two B vitamins — niacin and riboflavin — help in this process.

Niacin and riboflavin form the carriers of H⁺ ions from the Krebs cycle to the electron transport system. Niacin forms *nicatinamide adenine dinucleotide (NAD)* and can combine with H⁺ to form NADH. Riboflavin forms *flavine adenine dinucleotide (FAD)* and combines with H⁺ in a similar way to form FADH.

The bottom line: B vitamins are very important in energy production, just not the way you may have thought.

Using glucose and fat to make ATP

As the earlier section "Cooking up ATP, oxygen free: Anaerobic glycolysis" explains, glucose can be broken down to two pyruvic acid molecules through the process of glycolysis. Pyruvate is a 3-carbon molecule that can be converted into a key molecule that the mitochondria can use to make energy when oxygen is present. That molecule is a 2-carbon unit called *acetyl CoA*. The leftover carbon is lost when carbon dioxide (CO_2) is formed.

2 pyruvate ——— 2 acetyl CoA + 2 CO_2 (produced as waste)

Fat is a very long chain of carbon, hydrogen, and oxygen. Fats can be brought to the mitochondria, and from there, a number of 2-carbon acetyl CoA units can be created in a process called *beta-oxidation* (in this context, *beta* refers to the number 2). Think of a fat like a long chain of sausage links, with each individual link being a 2-carbon beta unit.

Both carbohydrates and fats can be broken down into acetyl CoA molecules, which are the primary entry points into the Krebs cycle. In this way, both glucose and fat can undergo aerobic metabolism. You can read about the Krebs cycle in the later section "The Krebs cycle: The body's hydrogen producer."

Using protein for ATP

Protein isn't a big contributor for ATP production. In fact, under normal conditions (like eating enough food each day), you may get only about 5 percent to 10 percent of your energy from protein. Instead, protein is used to make things: muscle, enzymes, and antibodies — pretty important stuff! Muscle is the greatest storage location for protein, so anything that steals protein results in the loss of muscle mass.

Protein can be used for energy in two ways:

- **As components within the Krebs cycle:** The big wheel of the H+ ion producer needs a number of molecules to keep things going. Protein can be broken down into its building blocks (called *amino acids),* which can then be used within the Krebs cycle (after the liver takes out the nitrogen, that is).

- **As energy after being converted into glucose:** If carbohydrates are low in the body, the liver can convert protein into glucose in a process called *gluconeogenesis* (think "glucose-new-make"). Although using protein in this way can save you when you don't have any carbs, it uses up your muscle, so it's not a good idea in the long run.

The Krebs cycle: The body's hydrogen producer

For your "battery" to run, it has to produce H^+ ions. Fortunately, that is what the Krebs cycle does. The Krebs cycle is a series of chemical reactions that rearrange molecules and release H^+ ions.

As Figure 4-6 shows, enzymes are needed for each reaction. For example, the starting molecule, acetyl CoA, is rearranged to form citrate, then isocitrate, then alpha-ketoglutarate. At that rearrangement, an H^+ ion is released and carried onward for ATP generation. The process continues with additional rearrangements (succinyl CoA → succinate, and so on).

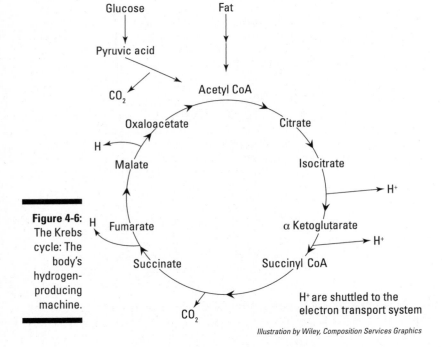

Figure 4-6: The Krebs cycle: The body's hydrogen-producing machine.

Illustration by Wiley, Composition Services Graphics

CO_2 is released during "turns" of the cycle as a waste product, and H^+ ions that are released are carried or shuttled onto the location for energy production (the electron transport system). As long as the big wheel keeps on turning, H^+ can be produced through the molecular rearrangements.

The electron transport system (ETS): Running the battery

The electron transport system (ETS) can be thought of as the guts of a battery. Located on the inner mitochondrial membrane, it's responsible for harvesting the energy from the H^+ ions that are produced by the Krebs cycle to be used for ATP production. Just like a battery gets its juice by having an electrical gradient, the ETS creates a chemical gradient across the inner mitochondrial membrane. This is where the energy for ATP production comes from.

Figure 4-7 illustrates this process. Electrons are removed from H^+ and passed down a series of reactions (almost like water running over a dam). The energy created by this cascade of electrons is what generates ATP. At the end of the sequence, H^+ remains and must be eliminated.

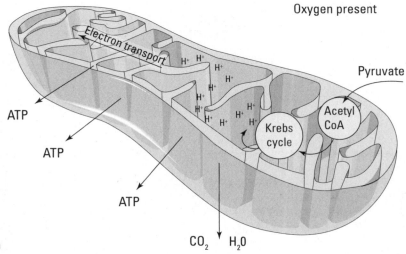

Figure 4-7:
The electron transport system (ETS): Using H^+ to make ATP.

Illustration by Wiley, Composition Services Graphics

1. **Due to the presence of oxygen, pyruvate gets broken down into acetyl CoA, which goes into the mitochondria.**

2. **The acetyl CoA goes through the Krebs cycle, which produces H^+ ions.**

3. **The H^+ ions are sent along the ETS, where the energy from the H^+ is used to drive ATP production.**

4. **The harvested energy is released as ATP, and the waste is removed as CO_2 and water.**

Water under the bridge: Understanding oxygen's role

Now you may be thinking, "But H⁺ ions cause fatigue! I don't want H⁺ accumulating in my mitochondria!" Ah, well, this is where oxygen plays its part in the story.

When you combine H⁺ and oxygen, you get water:

$$2H^+ + O = H_2O$$

Water is as neutral a substance as it gets. As long as oxygen is available, the H⁺ is neutralized. Translation: The oxygen acts as a buffer, eliminating the H⁺ ions and assisting with ATP production. If you want to make a lot of ATP, you need to get a lot of oxygen into the mitochondria. Therefore, oxygen use is related directly to the amount of work you do aerobically (and it's a nice thing to measure, which you discover later in the next section).

Measuring Metabolism: VO₂

As cool as it would be to have a camera small enough to watch what goes on inside a cell, that's not possible currently. To get an indication of how hard cells are working, therefore, we are left to measure whatever we can. In terms of metabolism, that means measuring either what is produced as a result of metabolism or what is used up.

Why no carbs means muscle loss

Glucose is essential to burn fat. Fat and glucose both enter into the Krebs cycle (that big wheel of H⁺ producers, explained in the section "The Krebs cycle: The body's hydrogen producer") at the same point, as acetyl CoA. However, glucose also contributes to other parts of the cycle. Without glucose, that big wheel won't keep on turning.

Without glucose, fat continues to accumulate in the form of acetyl CoA but can't get through the Krebs cycle. The body responds by letting the liver convert the acetyl CoA to an acid called a *ketone*. Ketones slow the metabolic rate, reduce appetite, and build up levels of acid in the blood (and your body does not like acid). Left unchecked, ketosis can lead a person with diabetes into a coma.

In healthy people, restricting carbs (as you do in a low-carb diet) also brings on ketosis. However, instead of just messing with their fat metabolism, they also start losing muscle mass. Here's why: Protein can be used for ATP production within the Krebs cycle, *and* the liver can turn protein into glucose! If you deprive your body of glucose, it responds by depriving you of your muscle mass, a situation that slows metabolic rate even more and makes it hard to lose body fat. So eat your carbs!

The calorie — a measurement of heat

Your system runs on ATP. The chemical energy from ATP breakdown is used to make your muscles move and to perform all other actions. But did you know that only about 20 percent of the ATP energy goes to your muscle? What happens to the other 80 percent? It produces heat! That's why shivering keeps you warm and why you need to sweat when you exercise. One way to determine work intensity and calories burned is to measure the amount of heat produced during exercise.

In the early days of measuring heat, exercisers were placed inside a sealed chamber surrounded by water; oxygen was fed into the chamber, and carbon dioxide was removed. As the person exercised, researchers measured the change in water temperature. A *calorie* was defined as the amount of heat it took to raise 1 milliliter (or 1 gram) of water 1 degree centigrade.

Because raising 1 milliliter 1 degree takes a pretty small amount of heat, the standard is to deal in numbers a thousand times bigger. So a *kilocalorie* (heat needed to raise 1 *liter* of water 1 degree centigrade) is what people usually mean when they talk about calories.

Measuring the volume of oxygen (VO_2) consumed

Measuring the heat produced during exercise is difficult. Fortunately, oxygen, which is used to make ATP, can also be measured. As a result, it has become a primary variable when determining both work intensity and calories burned.

Oxygen concentrations in air can be used to measure heat. The air around you is about 21 percent oxygen. Now take a breath and then exhale. The concentration of oxygen in the air you just exhaled is only about 16 percent oxygen. Where did the rest of the oxygen go? You used it in the mitochondria to make ATP!

By measuring how much air you are moving each minute (breathing in and out), you can then determine the *volume of oxygen* (VO_2) you use each minute. VO_2 is usually measured in liters used per minute (L/min). At rest, VO_2 may be as low as 0.25 L/min; during peak exercise, it may be as high as 6.0 L/min.

If you worked as hard as you could, perhaps running faster and faster on a treadmill until you fatigued, eventually you will hit the peak of your ability to use oxygen to make ATP. This is known as the *maximal oxygen uptake,* or VO_2max.

Because the use of oxygen is directly related to physical work, measuring VO_2 is the primary tool to assess how hard the work is. VO_2 is directly linked to ATP productions and heat produced (calories); therefore, if you know VO_2, you can estimate calories burned. For every liter of oxygen consumed, about 5 kilocalories are burned. So the equation is

calories burned = VO_2 L/min × 5

Exercise physiologists use this equation to determine how many calories various activities burn.

Comparing fitness levels: VO_2 and body weight

Suppose you have a 200-pound man and a 120-pound woman. Each can lift 150 pounds. Who is stronger? Trick question? Maybe.

On one hand, both have the same strength (each can lift 150 pounds). But based on size, the woman is stronger. Factoring strength based on body weight seems like a better way to compare.

When you want to compare the fitness levels of individuals, you need to determine VO_2 according to body weight. Follow these steps:

1. **Convert each person's weight into metric units.**

   To convert pounds into kilograms, you take the weight in pounds and divide it by 2.2.

2. **Covert L/min to ml/min by multiplying the L/min value by 1,000.**

3. **Adjust for body weight.**

   You divide the ml/min value by the weight in kilograms. This calculation gives you the milliliters of oxygen used per kilogram of body weight each minute, a unit represented as ml/kg/min.

4. **Compare the VO_2max to determine who is more fit.**

   As Table 4-1 shows, the higher the VO_2max number, the more fit the individual.

Say you're comparing a man who weighs 180 pounds and a woman who weighs 135 pounds. Both have a VO_2max of 4.0 L/min and you want to determine who is more fit. Table 4-2 shows the comparison, using the preceding steps.

Table 4-1	Fitness Categories Based on VO_2max (in ml/kg/min)		
	Low Fitness	*Average Fitness*	*High Fitness*
Men	< 45	46–55	56+
Women	< 40	41–49	50+

Table 4-2	Comparing VO_2max: An Example	
	Man	*Woman*
Weight converted from pounds to kilograms	180 lbs ÷ 2.2 = 81.8 kg	135 lbs ÷ 2.2 = 61.4 kg
VO_2max converted from L/min to ml/min	4.0 L/min × 1,000 = 4,000 ml/min	4.0 L/min × 1,000 = 4,000 ml/min
ml/min adjusted for body weight	4,000 ÷ 81.8 = 48.9 ml/kg/min	4,000 ÷ 61.4 = 65.1 ml/kg/min
Conclusion	His VO_2max is 48.9 milliliters of oxygen used per kilogram body weight each minute. The man's fitness level is average.	Her VO_2max is 65.1 milliliters of oxygen used per kilogram body weight each minute. The woman's fitness level is in the highest fitness category!

By using this method of displaying VO_2max, you can compare people of different sizes and make tables of fitness norms (so you can compete against your friends or at least see whether you have a decent level of fitness!)

Measuring metabolism during exercise

Technology has advanced to the point where it's now possible to measure the concentration of oxygen and carbon dioxide in every breath taken. All you have to do is capture the air and redirect it to an analyzer, like the one shown in Figure 4-8.

Just as items in a store cost a certain amount of money, exercise and work intensity cost a certain amount of oxygen to make ATP.

Here are some basic principles of VO_2 related to work:

- ✔ If work intensity goes up, VO_2 goes up.
- ✔ Whoever has a higher VO_2max can do more work.
- ✔ Work that uses a lot of muscle mass uses more oxygen (VO_2) and, therefore, burns more calories.

Given this information, which activity do you think burns more calories: sit-ups or walking? You've probably heard that if you want to lose fat around the belly, you should do sit-ups. But how much work does a sit-up entail versus walking? If you consider that sit-ups use very little muscle (abdominals and hip flexor muscles) and walking uses large muscles (legs, hips, and arms), you can see that walking does much more work, so it "costs" more ATP and more oxygen, and it burns more calories.

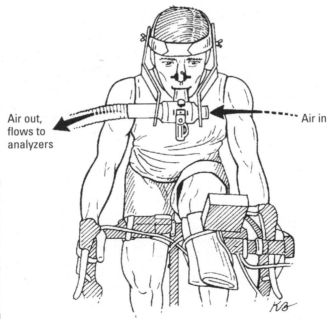

Air out, flows to analyzers

Air in

Figure 4-8: Measuring VO_2 by analyzing the concentration of gasses in your breath.

Illustration by Kathryn Born, MA

Other activities that use a lot of muscle and, therefore, use a lot of calories include biking, cross-country skiing, dancing, hiking, jogging, rowing, swimming, and walking up a hill (which uses even more leg muscles than regular walking).

Measuring changes in metabolism: The anaerobic threshold

In earlier sections, we explain how both aerobic and anaerobic systems make ATP. The aerobic systems, using fat, carbohydrates, and oxygen, can keep making ATP. However, these systems work only at low to moderate work intensity; once you go beyond that, you have to start also using the anaerobic system. But anaerobic metabolism has a key downside: lactic acid, which leads to fatigue. As a person moves from easy to heavy work, a transition in metabolism occurs that shows up as a big increase in lactic acid in the blood. This sudden increase in acid represents the *anaerobic threshold (AT)*.

That's one smart treadmill!

As we explain in this chapter, for every 1 liter of oxygen used, about 5 kilocalories are burned. By knowing VO_2, you can estimate the number of calories burned. Researchers have established prediction equations for walking, running, cycling, and stepping that can estimate the VO_2 cost of work intensity. These equations are built into exercise equipment software. All the equipment needs is the workload and your body weight to determine how many calories you burn. Here is an example:

Julie is walking on a treadmill at 3.5 mph (93.8 meters/min). She weighs 125 pounds (56.8 kg). The equation for VO_2 estimation in ml/kg/min for level walking is

$$VO_2 \text{ ml/kg/min} = \left(\text{speed in meters/min} \times 0.1 \right) + 3.5$$

Plugging in Julie's numbers, you get

$$VO_2 \text{ ml/kg/min} = \left(93.8 \times 0.1 \right) + 3.5$$
$$= 9.4 + 3.5$$
$$= 12.9$$

Next you convert VO_2 to L/min by multiplying Julie's VO_2 ml/kg/min by weight in kilograms and then dividing the result by 1,000:

$$12.9 \text{ ml/kg/min} \times 56.8 \text{ kg} = 732.7 \text{ ml/min} \div 1,000$$
$$= 0.733 \text{ L/min}$$

Finally, to get the kilocalories, you multiply VO_2 L/min by 5 (every L/min of oxygen used burns 5 kilocalories):

$$\text{kilocalories} = 0.733 \text{ L/min} \times 5$$
$$= 3.7 \text{ kilocalories per minute}$$

If Julie walks for 20 minutes, she burns 74 kilocalories ($3.7 \times 20 = 74$)

By inputting your body weight, exercise machines can give a pretty decent estimate of your calories burned. Pretty handy!

Think about what happens at low, moderate, and high intensity exercise:

At low-intensity

- ✔ Aerobic (slow twitch) muscle fibers are doing most of the work. These fibers do not produce lactic acid. In fact, they can even use lactic acid for energy!

- ✔ Aerobic metabolism (the Krebs cycle) is producing most of the ATP. With oxygen available, only water and carbon dioxide are produced.

- ✔ Fat and carbohydrates are both being used for energy.

At moderate intensity

- ✔ The work is getting harder, and oxygen delivery to the muscle can't quite keep pace with the exercise intensity.

- ✔ Fast twitch (anaerobic) muscles are starting to be used in addition to slow twitch muscles. Fast twitch fibers make lactic acid, so lactic acid starts to accumulate.

- ✔ Fat is used less as a fuel, because more anaerobic work is being done, and carbs are used more and more.

At high intensity

- ✔ All your muscle fibers are being used, and lots of lactic acid is produced!

- ✔ With anaerobic metabolism kicked into high gear, lots of lactic acid is produced!

Because lactic acid is related to fatigue, the point where lactic acid increases is a key index of how hard you can work before fatigue (see Figure 4-9). For this reason, athletes tend to race at an intensity just below their anaerobic threshold.

Measuring lactic acid (which you do by taking a blood sample) is not the only way to measure the anaerobic threshold. As we note earlier, lactic acid can be converted to CO_2 and then breathed off (ventilation). Researchers can actually measure either CO_2 production or even changes in ventilation to identify the threshold — tactics that are easier than poking a finger to get a blood sample!

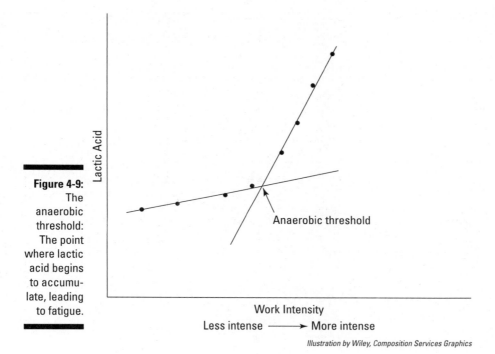

Figure 4-9: The anaerobic threshold: The point where lactic acid begins to accumulate, leading to fatigue.

Illustration by Wiley, Composition Services Graphics

Training for Improved Metabolism: It's the Enzymes!

When it comes to making ATP, you know that you need a starting fuel (phosphocreatine, carbohydrates, or fat) and then a way to transform the molecules to get the energy out of them. So what magic ingredient transforms the molecules? Enzymes! As we explain earlier, enzymes are the components that drive chemical reactions. What's particularly interesting — and helpful in the context of improving performance — is that you can create more enzymes. How? By training! The key is to tailor the training needs to the type of enzyme changes you want.

Predicting 10K time, using the anaerobic threshold

Imagine you just tested an athlete on a treadmill doing a running test. You measured ventilation during the test. Can you predict the running speed of the athlete during a 10K race? You betcha. Take a look at illustration below.

Notice that, at 8.0 mph, the ventilation suddenly increases at a higher rate. That is the point at which lactic acid is accumulating, carbon dioxide is accumulating, and ventilation is increasing.

Running at 8.0 mph, your runner is running at a 7.5 minute-per-mile pace (60 min/hr ÷ 8.0 mile/hour = 7.5). Because you know that a 10K race

is 6.2 miles, you can do the calculation to estimate how long it will take this runner to complete a 10K race:

6.2 miles × 7.5 minutes each mile = 46.5 minutes (46 minutes, 30 seconds)

So the 10K time is closely related to the pace at which the anaerobic threshold occurs. You can use these kinds of calculations to track training progress or to even compare runners, because any changes in the anaerobic threshold mean changes in race pace or performance. Useful!

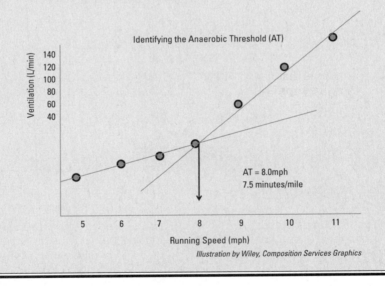

Identifying the Anaerobic Threshold (AT)

AT = 8.0mph
7.5 minutes/mile

Illustration by Wiley, Composition Services Graphics

Getting better at what you're doing: Training specificity

They say practice makes perfect. Well, that may be true only if the practice itself is near perfect. Bodies tend to adapt to the exercise conditions they're given. Run slowly, and you improve at running slowly. *Specificity of training* means that improvements in training are specific to what you are doing and includes things like the speed of movement, the load lifted, the intensity of the exercise, the angle or position of the limbs, the environment . . . you get the idea. Improvement in training is very finicky!

Energy systems are simply a combination of chemical reactions. They start with a molecule (phosphocreatine, glucose, or fat) and use enzymes to help the reactions take place. The more enzymes you have, the faster you make ATP, and the way to get more enzymes is to train in ways that use particular enzymes.

Training the ATP-phosphocreatine system

The phosphocreatine system (explained earlier in "Phosphocreatine: An Immediate Source of ATP") provides the energy for the highest intensity of activity. You only have up to about ten seconds of phosphocreatine stored. A training regimen for improving this system includes these characteristics:

- High-intensity exercise intervals of five to ten seconds in length
- Short rest bouts between intervals (don't worry, no lactic acid was made during this activity)
- Activities that use the same muscle you would use for competition because improvements are specific to the muscle used

By following such a regimen, you should see these training improvements:

- Phosphocreatine breaking down at a faster rate
- Some increase in phosphocreatine storage (although diet influences this, too)
- Improvement in all-out effort because you make energy faster

Training the anaerobic glycolytic system

The anaerobic glycolytic system (see "Anaerobic Glycolysis: Fast Energy with a Price") uses glucose as its primary fuel. It's useful for high-intensity activity that lasts longer than ten seconds and that can last as long as four minutes. The side effect of this system is lactic acid.

If you want to improve this system, do the following:

- Train, using a high-intensity interval activity of between 30 and 60 seconds.
- Include rest intervals of between one and five minutes of light activity (which gets rid of lactic acid). Then hit it again!
- Use the muscles you'll use for competition.

This training regimen uses up glycogen, so you may need a rest day to restore.

Following are the training improvements you should see:

- Faster ATP production (more enzymes)
- Faster removal of lactic acid (thank goodness!)
- Faster speed and quicker recovery — in other words, improved performance

Training the oxidative (aerobic) system

The oxidative system (explained in "The Oxidative (Aerobic) System: It Just Keeps Going and Going") is the ultimate ATP producer, using fats, carbs, and even proteins to make ATP. The aerobic system can be trained a number of ways, depending on what you're trying to achieve:

- If you are in lousy shape to begin with, you can improve your aerobic fitness by exercising lightly as few as two times per week.
- If you want to train the aerobic system, engage in moderate exercise, which is a great because it uses mainly the aerobic system to make ATP.
- If you want to improve the aerobic system, engage in high-intensity training. You may even see improvements in the aerobic system while working on the glycolytic system. Bonus!

Use large muscle groups so that you maximize the use of the system. Activities like walking, jogging, swimming, hiking, biking, and rowing are ideal.

You should see these training improvements:

- More enzymes, producing more ATP
- Less lactic acid at the same running speed
- Improved ability to get oxygen into the mitochondria
- More mitochondria, which is like having a bigger aerobic muscle
- Improved ability to deliver blood (which carries oxygen) to the muscle
- A bigger heart, which means a bigger pump to push more blood and more oxygen through your body

If you want to make energy both during very high-intensity activity and more modest intensity, you need to train specifically for it. Your body is like putty, but you have to do some work to shape it the way you want, both inside and out!

Chapter 5

The Body's Engine: The Cardiovascular System

. . . .

In This Chapter

▶ Discovering the function of the heart and circulatory systems

▶ Linking key cardiovascular variables and exercise

▶ Noting how the cardiovascular system adapts as a result of exercise training

. . . .

All the contracting muscles, nerve stimuli, and metabolic activity of the body would not take place if the necessary nutrients were not made available. The way things like glucose, fat, protein, and oxygen get transported to the cells is through blood flowing to the body's tissues. Wastes, like carbon dioxide, are also removed through blood flow.

The heart is the pump that moves the blood through the blood vessels throughout the body. If this pump is strong, you can do a lot of work. If it weakens, you weaken. In this chapter, we look at how the cardiovascular system works at rest and during exercise.

The Heart's Structure: A Muscle Made to Pump

The heart is designed to be an efficient pump. Its structure and the way it works all serve to move oxygenated blood to all the tissues and organs in your body. Becoming familiar with the different components that make up the heart helps you understand how the heart works and why you can't talk about movement or exercise without a firm knowledge of the heart muscle.

Heart chambers and valves

The heart collects blood and, like a policeman directing traffic, sends it out to two different locations. It has two different types of chambers:

- ✔ **Atria:** The atria are small chambers on the topside of the heart. They collect blood that is returning to the heart via the veins.

- ✔ **Ventricles:** The ventricles are the larger, more muscular chambers on the lower half of the heart. They get their blood from the atria above and later squeeze to pump the blood away from the heart.

Heart valves, which open and close in a coordinated way to help keep the blood moving in one direction, are between the atria and ventricles, as well as between the blood vessels and the heart.

Two halves of the whole

Just as the atria on the top portion and the ventricles on the lower part of the heart have their own purpose, the left and right sides of the heart (each with one atrium and one ventricle) have different functions:

- ✔ **The right side:** The right side of the heart is responsible for two things:

    - Collecting all the blood that has been out delivering oxygen to the muscles and tissues. Blood returns to the right-side atrium by way of large veins called *vena cavae* (singular *vena cava*).

    - Pumping the blood out of the right-side ventricle and to the lungs, where it can pick up oxygen.

- ✔ **The left side:** The left side of the heart works in the same way as the right side, except for the direction of the blood flow:

    - The left-side atrium collects oxygen-rich blood from the lungs by way of the pulmonary veins. This blood is ready to feed oxygen to the tissues.

    - The left-side ventricle pumps the oxygenated blood out to the entire circulatory system of the body.

When the left ventricle squeezes blood out, you can feel the wave of blood. Place your fingers on the side of your neck (carotid artery) and you can feel the waves go by. This is one way to measure heart rate!

Seeing How the Heart Works

Imagine that you're playing in a swimming pool, and you decide to squirt water with your hands. If you're like most people, you'll cup your hands to make a chamber, fill the chamber with water, and then squeeze down quickly to force the water out of the opening you leave. The heart works in a similar way. It moves blood throughout the body by first filling its chambers and then squeezing down to force blood through the cardiovascular system.

Watching the blood flow through the heart

While the right side of the heart collects oxygen-depleted blood from the body and sends it to the lungs for oxygen, and the left side of the heart collects oxygenated blood from the lungs and sends it out to the body (refer to the earlier section "Two halves of the whole"), their actions occur simultaneously. This is why the heart beats in a nice, coordinated fashion.

Table 5-1 outlines the steps involved in the movement of blood through the heart. You can see this path in Figure 5-1.

Table 5-1		How Blood Moves through the Heart
	Right Side of the Heart	**Left Side of the Heart**
Step 1	Deoxygenated blood returns to the right atrium, where it collects. This is known as the *venous return*.	Oxygen-rich blood returns to the left atrium from the lungs, where it just picked up a fresh supply of oxygen.
Step 2	The valve separating the right atrium and the right ventricle opens, and two-thirds of the blood flows into the ventricle. The remaining one-third stays in the atrium until the atrial muscle contracts, forcing it out. *Tip:* Think of this action as akin to wringing the water out of a sponge: When you take a sponge out of a water bucket, a lot of the water just drips off the sponge, but to get the rest, you have to give the sponge a squeeze.	The valve separating the left atrium and the left ventricle opens, and two-thirds of the blood flows into the ventricle. The remaining one-third stays in the atrium until the atrial muscle contracts, forcing it out.

(continued)

Table 5-1 *(continued)*

		Right Side of the Heart	**Left Side of the Heart**
Step 3		The right ventricle contracts, squeezing down on the blood that's inside, and the pressure builds. The valve between the right atrium and right ventricle closes to keep the blood from going backward. *Lup!* You can hear the sound of the closing with a stethoscope.	The left ventricle contracts, squeezing the blood, and the pressure builds. The valve between the left-side atrium and ventricle closes, keeping the blood from backing into the atrium. *Lup!* You can hear the sound of the closing with a stethoscope.
Step 4		As the right ventricle squeezes and the pressure builds, the valve holding the blood in the right ventricle opens, and the blood is pushed out to the lungs through the aorta.	As the left ventricle contracts, the valve holding the blood in the left ventricle opens, and the blood is pushed out to the entire body through the aorta.
Step 5		The heart muscle relaxes. Some of the blood that was pumped out flows backward to the heart and closes the heart valves. *DUP!* You can hear the sound of the closing with a stethoscope.	The heart muscle relaxes. Some of the blood that was pumped out flows backward to the heart and closes the heart valves. *DUP!* You can hear the sound of the closing with a stethoscope.

The sounds your heart makes are the product of the valves closing in Steps 3 and 5. In Step 3, when the valves between the atria and ventricles close to stop the blood from flowing back into the atria, you hear the first beat. In Step 5, when the valves in the ventricles close to stop the blood from flowing back into the heart, you hear the second beat. Because the valves leading from the ventricles to the lungs and body are larger than the valves separating the atria and ventricles, they produce a louder sound when they close: lup DUP!

The term used to denote contraction of the ventricle (Step 4) is *systole* (pronounced "SIS-tol-ee"). The term used to denote relaxation of the ventricle (Step 5) is *diastole* (pronounced "die-ASS-tol-ee").

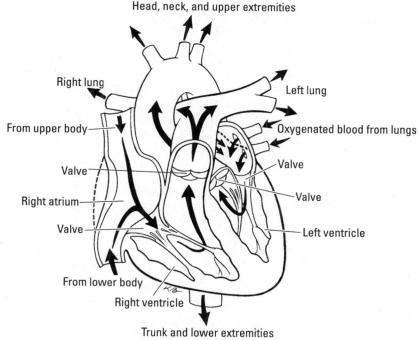

Head, neck, and upper extremities

Right lung

Left lung

From upper body

Oxygenated blood from lungs

Valve

Valve

Valve

Right atrium

Valve

Left ventricle

From lower body

Right ventricle

Trunk and lower extremities

Illustration by Kathryn Born, MA

Figure 5-1:
The direction of blood flow through the heart.

A noisy heart: Heart murmurs

Normally the sounds you hear during the pumping cycle of the heart are clear and sharp (lup DUP. . . lup DUP. . . lup DUP). But, sometimes heart valves get leaky, either because they don't close all the way or because the seals leak. The result is that blood squirts backward through the gap and makes a rumbly sound.

Leaky valves between the atria and ventricles would cause a sound like "phyyyth dup . . . phyyyth dup . . . phyyyth dup. Heart murmurs can be minor and no worry, or they can be more significant and interfere with the heart's ability to adequately pump blood.

Getting blood to the heart

The heart is very generous, sending oxygen-rich blood throughout the body. But what about the heart muscle itself? The heart is pumping and pushing and using quite a bit of oxygen, so it needs a constant supply of oxygen, too. Figure 5-2 shows the coronary arteries (darkened in the image). Notice that they originate in the aorta, just past the valves of the heart.

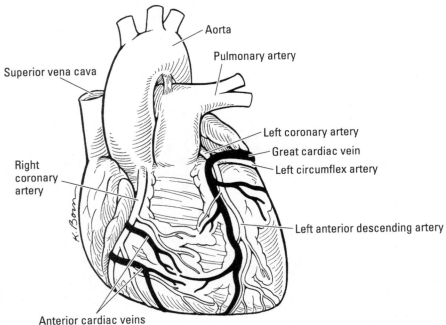

Aorta

Pulmonary artery

Superior vena cava

Left coronary artery

Great cardiac vein

Left circumflex artery

Right coronary artery

Left anterior descending artery

Figure 5-2:
The coronary arteries originate at the aorta.

Anterior cardiac veins

Illustration by Kathryn Born, MA

When the heart is in systole (contraction), the arteries that supply blood to the heart squeeze shut, meaning the heart is doing work but not getting any oxygen. After systole is completed and some blood rushes back toward the heart, the ventricle is in a relaxed state (diastole). During this phase, the blood coming back to the heart enters the coronary arteries, which are located on the aorta just beyond the valves between the aorta and the ventricle (see Figure 5-3).

The heart uses oxygen during the contraction phase and receives its supply of oxygen during the relaxation phase.

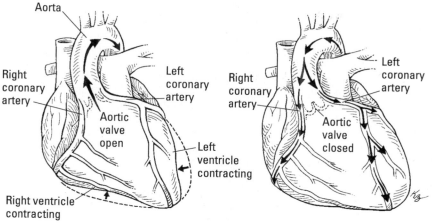

Figure 5-3: During diastole, the blood flows back toward the heart, providing it with oxygen.

Systole

When the ventricles contract, the walls squeeze the coronary arteries shut, preventing blood flow. The blood within the ventricle is pushed out through the aorta.

Diastole

When the ventricles relax, suction pulls blood back toward the heart. The aortic valve closes, and backward pressure sends blood into the coronary arteries.

Illustration by Kathryn Born, MA

Identifying the force behind the heart beat: Blood pressure

If you've ever pumped up an air mattress with a foot pump, you know that you have to push hard enough to overcome the air pressure that is already in the mattress in order to push air into the mattress. The heart functions in a similar manner: It has to generate enough pressure to push blood out into the system of arteries that deliver blood (and oxygen) to the tissues. It uses two primary pressures, systolic and diastolic blood pressure, to accomplish this task (see Figure 5-4):

✔ **Systolic pressure:** As we note in the earlier section "Getting blood to the heart," systolic blood pressure is the pressure generated during the ventricle's contraction phase. Imagine that you are holding one end of a long rug and you snap it. You see a wave of rug move away from you. In a similar fashion, when the ventricle contracts, a large wave of blood is sent away from the heart. This wave actually stretches the arteries (they bulge as the wave moves past). Normal values for systolic blood pressure at rest range from 90 to 120 mmHg. (**Note:** Blood pressure is reported in *millimeters of mercury,* abbreviated to *mmHg.*)

Try feeling this systolic pressure wave yourself. Place the pads of your first two fingers across either your carotid artery (at the neck) or your radial artery (palm side up, thumb side of your wrist). If you get the location right, you can feel the ventricular pulses as they move past.

✔ **Diastolic pressure:** Because the vessels are full of blood, pressure already exists in the system. The pressure in the circulatory system during the resting phase is called the *diastolic blood pressure*. Normal values for diastolic blood pressure range from 50 to 80 mmHg.

Figure 5-4:
The contraction phase generates systolic pressure; the resting phase, diastolic pressure.

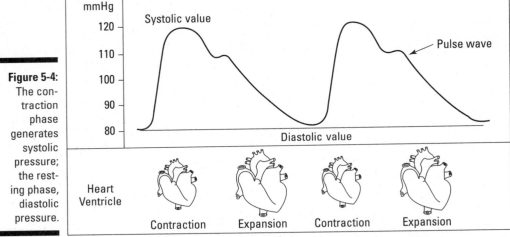

Illustration by Wiley, Composition Services Graphics

Setting the pace: What controls heart rate?

A single cardiac cell is a wondrous thing. All by itself, it'll beat in a nice rhythmic fashion. Of course, if all cardiac cells decided to beat at their own pace, the heart would never be able to squeeze in a coordinated movement. Therefore, special heart cells tend to pace the entire heart so that blood moves through in a coordinated fashion.

Introducing the SA node

Within the atria and ventricles is a specialized tissue that spreads across the chambers. This tissue receives an impulse and quickly spreads it to all the cardiac muscles in the chamber. All it needs is an initial pulse, which comes from one spot in the right atrium: the *sinoatrial node* (*SA node*).

Under pressure: Hypertension and heart disease

Blood vessels are fragile things. They stretch and can handle large loads of blood, but if they stay under pressure for too long, they begin to become damaged. Prolonged systolic pressure over 120 mmHg and/or diastolic pressure over 80 mmHg may cause the arterial walls to begin to scar and fill in. This narrows the arteries and is a key part of atherosclerosis, the leading cause of stroke and heart attacks. Because you can't feel high blood pressure (hypertension), you must have your blood pressure measured to discover whether you have a problem.

Pressure can hurt the heart as well. During systole, if the heart has to push against a lot of pressure (called *afterload*) or if it has a lot of blood returning to it (called *preload*), it must use more oxygen to do more work — a situation that can fatigue the heart and weaken it.

Think of the SA node as the heart's pacemaker. Although all cardiac tissue can contract on its own, the SA node seems to contract faster than all other heart tissue. Because it beats the other tissue to the punch, your heart rate is set according to the rate at which the SA node fires.

Stimulating and contracting the heart, step by step

The sequence of stimulus and contraction is like a carefully choreographed dance that moves blood through the heart (see Figure 5-5):

1. **When the atria have filled with blood and the blood begins to flow to the ventricles, the SA node fires.**

2. **This electrical signal sweeps across the atria, stimulating the atria to contract and push the blood into the ventricles.**

3. **The electrical signal is delayed at a junction between the atria and the ventricles, called the *atrioventricular node (AV node)*.**

   The delay at the AV node lasts only about one-tenth of a second, but it gives the atria a moment to do their work. You don't want the ventricles to contract while the atria are contracting. If they did, the blood would go forward and backward.

4. **The electrical signal emerges from the AV node and sweeps across the ventricles through fast conducting tissues (left and right *bundle branches* and *purkinje fibers*), causing an almost immediate stimulation of the ventricle, followed by ventricular contraction.**

5. **The stimulated cells reset (*repolarize*) to their original state.**

   This happens in between heart beats.

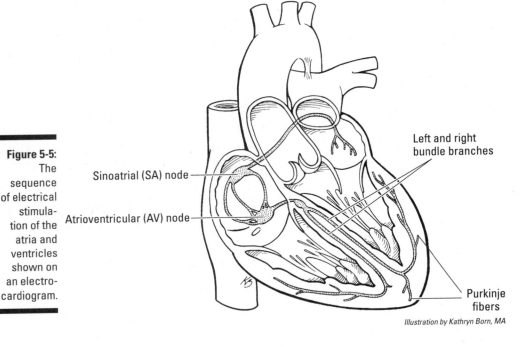

Figure 5-5:
The sequence of electrical stimulation of the atria and ventricles shown on an electrocardiogram.

Left and right bundle branches

Sinoatrial (SA) node

Atrioventricular (AV) node

Purkinje fibers

Fast or slow, what makes it go? The nervous system's influence on heart rate

If the SA node were allowed to pace heart rate without any influences, we would all have a resting heart rate of about 90 beats per minute. Is your resting heart rate that high? Probably not. So something is slowing it down. And something speeds it up when, for example, you start to exercise. That something is actually two things: your nervous system and hormones that can influence heart rate. Here's what happens during periods of rest and activity:

- ✓ **Your heart rate at rest:** During resting conditions, your body focuses on things like digestion and, well, rest! Under these conditions, the *parasympathetic nervous system (PNS)* is at work. The PNS has nerves that connect to the SA node and cause it to slow down at rest.

- ✓ **Your heart rate when action is required:** When you start to exercise or when you're under stress of some type, your body needs to push blood and oxygen to the muscle. In this case, the *sympathetic nervous system (SNS)* gets to work. The SNS also connects to the SA node, but when it kicks in, heart rate begins to rise in relation to the amount of stimulation. Some heart rates can go as high as 200 beats per minute!

The SNS also stimulates the adrenal gland, which contains a hormone called *epinephrine*. Epinephrine, when released into the blood, causes heart rate to accelerate. It's often referred to as the "fight or flight" hormone, because it's released under conditions of stress and helps to prepare the body for doing work.

So what happens when exercise begins?

1. **The PNS starts to shut down (well, you aren't at rest anymore, are you?), causing the heart rate to rise.**

2. **The SNS starts to kick in, and the heart rate continues to rise; systolic blood pressure also rises as the heart begins to push blood.**

   You are ready for some work!

3. **Epinephrine is released, your *bronchioles* (passageways in the lungs) dilate, and glucose and fat are dumped into your blood to send fuel to your cells.**

What about when exercise ends?

1. **The PNS comes online again, slowing the heart rate down.**

2. **The SNS starts to shut down, further reducing heart rate.**

3. **Epinephrine levels start to drop, as the release of epinephrine falls and the remaining epinephrine is broken down.**

Get moving to lower your resting heart rate!

Who has a lower resting heart rate, a trained aerobic athlete or an out-of-shape couch potato? If you guessed aerobic athlete, you are right! But why? One reason is the nervous system. Aerobic exercise, like walking, jogging, and biking, can actually train the PNS to be even stronger at rest, meaning that the resting heart rate goes even lower. A sedentary person may have a resting heart rate of 70, but an aerobic athlete's resting heart rate could be as low as 40! In addition, the body can be trained to be less sensitive to the sympathetic nervous system, which may mean that heart rate won't go up so much when you experience stress. This phenomenon is one reason why exercise is a good stress reducer.

Key measures of heart function

Because your body depends on the blood and oxygen that your heart pumps out each minute, how strong your heart is is a key measure of exercise ability. Following are a few important components related to heart function:

- ✔ **Stroke volume:** The ventricle chamber holds the blood just before it's pumped out for circulation. The more you can pump, the more blood and oxygen you can deliver. *Stroke volume* refers to the amount of blood that is pumped out of the ventricle with each heartbeat. Usually stroke volume is about 70 milliliters per beat in a resting heart.

- ✔ **Ejection fraction:** Ejection fraction is very similar to stroke volume, but it's a better indicator of heart strength. *Ejection fraction* refers to the percentage of blood that is pumped with each heartbeat. A normal, strong heart can eject about 60 percent of its blood with every beat, whereas a weak heart ejects less than 50 percent.

- ✔ **Rate pressure product (RPP):** RPP is a key indicator of the oxygen demand of the heart muscle. Because the heart needs oxygen to push, how hard it pushes (systolic blood pressure) and how fast it pushes (heart rate) influence oxygen need. The harder and faster the heart pushes, the more oxygen it needs. You use the following formula to calculate RPP:

  RPP = heart rate × systolic blood pressure

- ✔ **Cardiac output:** The total "horsepower of the heart" is related to how much the heart can pump with each beat, as well as how fast the heart is beating. *Cardiac output* is the total amount of blood that is pumped by the heart each minute. You can calculate it by using this simple equation:

  cardiac output = heart rate × stroke volume

Cardiac output is directly related to work intensity. If work intensity goes up, cardiac output goes up.

Delivering Fresh Air to Your Cells

As we explain in Chapter 4, producing energy (ATP) requires oxygen to help run chemical reactions in the mitochondria. Although oxygen is certainly available in the atmosphere, how does oxygen get from the atmosphere into your lungs and then into your blood and finally into your cells? Through a series of steps and biological processes, of course. Read on to discover how pressure, the simple act of breathing, and key biological processes let your body get all the oxygen it needs while also removing carbon dioxide.

Transporting oxygen through the body: The pressure gradient at work

A gas (like air) always moves from an area of higher pressure to an area of lower pressure. Think of a balloon. If you blow it up and let it go without tying off the end, what happens? The air comes rushing out. Why? Because of the difference in pressure between the air inside the balloon and the air outside the balloon! Without this pressure gradient, air and oxygen would not be able to travel through your body.

You may take breathing in and out for granted, but it's an essential part of getting oxygen to your cells. To breathe in and out, your body creates changes in air pressure:

- **Breathing in:** Two sets of muscles help you create a low pressure condition in your lungs. The *diaphragm* is a muscle just beneath your lungs that contracts during inspiration (breathing in). By contracting, the diaphragm creates low pressure in the chest, and this low pressure draws air in. The muscles between your ribs (the *intercostal muscles)* can also contract and expand your chest.

  Try this: Put your hand just below your chest and above your belly. Now breathe in through your nose (like you're smelling a flower). Do you feel your belly push outward a bit? This is your diaphragm contracting. Now take a big breath and breathe in hard. Feel your entire chest expand? Those are the intercostal muscles at work.

- **Breathing out:** After breathing in, you can simply relax the muscles you used to pull air in. When you do so, the pressure in your lungs rises and pushes the air back out. Or you can force the air out by contracting your intercostal muscles, which pulls the chest inward and sends the air out.

Paying attention to partial pressure

Imagine a room full of air. You can't see the air, but molecules of the gases that make up air (oxygen, carbon dioxide, and nitrogen) are bouncing all around in the room. Right now you're sitting at the bottom of a giant ocean of air (the atmosphere) that's pushing down on you. The amount of pressure the atmosphere generates is known as *barometric pressure*. You measure barometric pressure by measuring the height of a column of mercury in millimeters (mmHg). At sea level, air pressure is around 760 mmHg.

Air is comprised of three main gases:

Gas	Its Percentage of Air
Oxygen	20.93%
Carbon dioxide	0.03%
Nitrogen	79.04%

Using this information, can you guess how much pressure the oxygen component generates? If 20.93 percent of air is oxygen, it generates 20.93 percent of the pressure. Knowing this, you can do a simple calculation to determine how much of the pressure is generated by the oxygen alone.

To calculate pressure of the oxygen component, you multiply the air pressure at sea level by the concentration of the gas (well, actually the concentration in decimal form, so 20.93 percent is divided by 100 to get 0.2093). The resulting number is called the *partial pressure* because it is only a portion of the total pressure. The partial pressure of oxygen is represented as PO_2. The partial pressure of carbon dioxide is represented as PCO_2.

Here is the calculation for PO_2:

$$PO_2 = 760 \text{ mmHg} \times 0.2093$$

$$PO_2 = 159.1 \text{ mmHg}$$

You use the same method to determine the partial pressure of any gas (like CO_2 or nitrogen).

Each gas moves independently from higher to lower pressure, based on its own partial pressure.

Tracking the movement of O_2 and CO_2

Aerobic metabolism (refer to Chapter 4) needs oxygen to get to the mitochondria, and then it needs the carbon dioxide to get out of the cells and into the atmosphere. How are these feats accomplished? By creating partial pressure gradients for each gas so that they move from higher to lower pressures. As Table 5-2 shows, the pressure does the pushing! The pressure to move oxygen comes from the atmosphere, and the pressure to move CO_2 comes from the production of CO_2 in the cells.

Table 5-2		Partial Pressure Gradient for Oxygen (PO_2) and Carbon Dioxide (PCO_2)				
Atmosphere		**Lung**		**Blood moving through the Lung**		**Cells and Mitochondria**
$PO_2 = 159.1$	→	$PO_2 = 104.0$	→	$PO_2 = 40.0$	→	$PO_2 = 2.0$
$PCO_2 = 0.3$	←	$PCO_2 = 40.0$	←	$PCO_2 = 46.0$	←	$PO_2 > 46.0$

Notice in which direction the gases move: The oxygen moves from the atmosphere to the lung to the blood moving through the lung and finally to the cells and mitochondria. The carbon dioxide moves from the cells and mitochondria to the blood moving through the lungs to the lungs and finally back out into the atmosphere. Just what you want!

Carrying gases in the blood

Gases can exist not only in the atmosphere but also in fluid environments. If you've ever opened a can of soda or poured champagne, you've observed the gas coming out of a fluid: Think of the bubbles rising in a glass of champagne! Dissolved gases within a fluid establish the partial pressure of the gas in that fluid.

Unfortunately, the fluid part of blood cannot carry enough oxygen to supply your needs. Nor can it help get rid of all the CO_2 that your cells create. To solve this dilemma, the body has a variety of means to transport oxygen and carbon dioxide through the blood. These lifeboats for oxygen and carbon dioxide transport are the keys to your survival.

Transporting oxygen

The primary carrier of oxygen in our blood is *hemoglobin* (Hb), an iron-containing protein within your red blood cells. Hemoglobin is the stuff that gives your blood the red color (it's the iron), and it carries 99 percent of all your oxygen.

Each gram of hemoglobin can carry four oxygen molecules (about 1.3 milliliters of oxygen). The more hemoglobin you have, the more oxygen you can carry in your blood. If you are iron-deficient, your hemoglobin levels drop, and you become *anemic,* a condition that reduces your ability to carry oxygen in the blood.

Figure 5-6 shows hemoglobin picking up oxygen from the lungs and carrying it through the blood and then dumping the oxygen into the tissue. Typically hemoglobin is saturated with oxygen after leaving the lungs, with 98 percent of hemoglobin bonded to oxygen. Saturation levels drop as the oxygen is deposited in the tissue.

Hemoglobin can be finicky. It can either carry or dump oxygen. Which it does depends upon three key conditions:

- **The partial pressure of oxygen in the vicinity:** If PO_2 is high, as it is in the lungs when the blood moves through, the hemoglobin picks up and carries the oxygen. If PO_2 is low, like it is deep in the muscle tissue and the cells, the hemoglobin lets go of the oxygen and dumps it in the place where it's needed, inside the cell.

- **Temperature:** Hemoglobin holds on to oxygen more when the temperature is low. During exercise, when the temperature of the tissue rises, hemoglobin dumps its oxygen much faster and, in doing so, makes an exercising muscle happy.

- **The pH level:** The level of acidity in the blood influences hemoglobin's carrying capacity. If the pH level is more basic (as it is during rest), hemoglobin hangs onto the oxygen, but when things get acidic (as they do during exercise), the hemoglobin dumps more oxygen to the tissue.

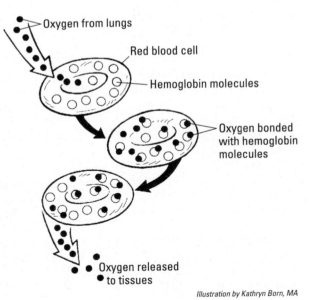

Figure 5-6: Red blood cells and the hemoglobin contained in them pick up, transport, and deposit oxygen.

Oxygen from lungs

Red blood cell

Hemoglobin molecules

Oxygen bonded with hemoglobin molecules

Oxygen released to tissues

Illustration by Kathryn Born, MA

Losing red blood cells . . . through your feet?

It may sound a bit odd, but runners can lose red blood cells through their feet. Think about the act of running: foot strike after foot strike. The loads on the feet can easily be six times your body weight. Now think of the small blood vessels passing through the tissue of the feet at the moment of each foot strike. The impact can be strong enough to actually destroy the red blood cells that are squeezing through the capillaries. Large-volume running can lower hemoglobin levels and cause anemia. Because running is an aerobic activity, the reduced oxygen carrying ability can really slow the runner down. For runners who have high training volumes, getting adequate iron in the diet is essential.

Following is what happens with hemoglobin and oxygen as they move through the body:

- **At the lung:** In the lung, hemoglobin encounters high PO_2, cooler temperatures, and a nice basic pH. Oxygen from the lung is pushed by a pressure gradient and picked up by hemoglobin. The oxygen-hemoglobin partnership, known as *oxyhemoglobin,* is secure. Off we go!

- **At the muscle:** Hemoglobin, carried in the blood, arrives at the muscle. Conditions in the muscle (especially if the muscle is exercising) are much different that conditions in the lung: low PO_2, an acidic environment, and a high temperature due to the ATP being used. Under these conditions, hemoglobin dumps its oxygen to the eagerly awaiting muscles (the oxygen deprived hemoglobin is called *deoxyhemoglobin*), and then it's back to the lung to repeat the process.

Aerobic performance enhancement: Doping

Red blood cells play a key role in aerobic performance, especially for endurance events, like the marathon, cross country skiing, or long cycling events. Sometimes athletes try to gain an advantage by boosting their red blood cell volume to unnaturally high levels. This is known as *blood doping.*

In blood doping, athletes often receive transfusions of red blood cells (often their own, which they have stored in a refrigerator) just before a big event. Because the blood carries more oxygen with more red blood cells, these athletes are able to work harder and perform better for up to 14 weeks after the infusion. Alternatively, a doping athlete can take a manufactured hormone called *erythropoietin (EPO),* which makes the body create more red blood cells. Both of these techniques can be tested for, and each year, athletes are disqualified for their cheating ways.

Transporting carbon dioxide

Carbon dioxide is the waste product of aerobic metabolism, and your body has to get rid of it! Once again, hemoglobin comes to the rescue! But CO_2 can also be carried in other ways:

- **Dissolved in the blood:** CO_2 dissolving in blood is just like gas dissolving in a soda! Only about 10 percent of CO_2 is transported in this way.

- **Bound to hemoglobin:** Rather than latching onto the iron portion of hemoglobin, CO_2 latches onto the protein portion of hemoglobin (together they are called *carbaminohemoglobin*). Twenty percent of CO_2 is carried this way.

- **Carried in the blood by altering its chemistry:** When PCO_2 is high, the carbon dioxide combines with water (with help from an enzyme in the red blood cell) to form a weak acid, called *carbonic acid:*

$$CO_2 + H_2O = H_2CO_3$$

Carbonic acid immediately breaks apart into two parts:

- **H^+ (hydrogen ion):** H^+ is like acid and can actually bind to your good friend hemoglobin and be carried away.

- **HCO_3 (bicarbonate):** Bicarbonate can actually be carried in the blood, so you can hold quite a bit of it. Seventy percent of your CO_2 is carried this way.

When the CO_2 reaches the lung, all the processes that have helped carry CO_2 reverse:

1. **Dissolved CO_2 diffuses into the lung.**

2. **Carbaminohemoglobin gives up its CO_2 to the lung.**

3. **Bicarbonate recombines with H^+, reforms carbonic acid, and the CO_2 then diffuses into the lung.**

Extracting oxygen from the blood: a-VO_2 difference

If you had a million dollars in a vault but didn't have the combination, the money really isn't going to help you, is it? Same with oxygen in the blood: Your oxygen-rich blood is great for your muscles to do work, but only if you can extract the oxygen!

Following are the keys steps to get the most oxygen out of the blood (see Figure 5-7):

1. **Starting with the artery, the hemoglobin loads up on oxygen.**

   Each hemoglobin molecule can carry 1.34 milliliters of oxygen. Hemoglobin values can range from 11 milligrams to 17 milligrams per 100 milliliters of blood. So, for example, if a hemoglobin level is 15, you can carry about 20 milliliters of oxygen ($15 \times 1.34 = 20$) in each 100 milliliters of blood.

2. **The blood vessels, and lots of them, irrigate the muscle.**

   The more blood vessels moving through the muscle, the closer the oxygen gets to the cells — much like irrigating a farm field! (You can read more about the role of the blood vessels in the next section, "Observing Blood Vessels in Action.")

3. **The partial pressure gradient draws oxygen from the hemoglobin and pushes it into the tissue.**

   The more mitochondria available (the mitochondria are organelles that produce ATP by using oxygen), the more oxygen extracted, and the more oxygen extracted, the more work you can do!

Figure 5-7:
The difference between the amount of oxygen in the arteries versus the amount in the veins represents your a-vO_2 difference.

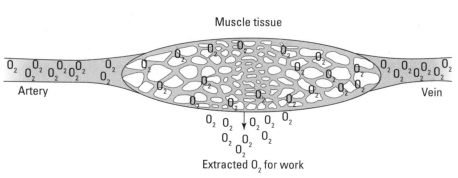

Illustration by Wiley, Composition Services Graphics

If you measure the amount of oxygen present in the arteries (Step 1) and then measure the amount of oxygen left over in the veins after it went through the tissue (Step 3), you can discover how much oxygen was extracted. Here is the calculation you use to determine *arterio-venous oxygen difference* (a-VO$_2$ difference):

$$\text{arterial } O_2 - \text{venous } O_2 = \text{amount of } O_2 \text{ extracted}$$
$$20 \text{ ml/dl} - 15 \text{ ml/dl} = 5 \text{ ml/dl}$$

So the a-vO$_2$ difference is 5 ml/dl.

As work intensity and the need for more oxygen increase, the a-vO$_2$ difference increases, meaning that more oxygen is extracted. Unfit people can't extract as much oxygen as fit people. They just don't have the "combination" to unlock all the available oxygen. That's why training is so important!

Observing Blood Vessels in Action

If you have summer water fights, you must learn proper water hose self-defense. If someone is chasing you around with a hose and squirting you with water, what do you do? You kink the hose to immediately stop the water flow. Then, when your friend looks at the nozzle to figure out why the water stopped, you unkink the hose, and he sprays himself! Fun right? In this way, you can control where and when the water flow occurs.

Your body is full of blood vessels that are squirting blood in many directions and to many places. Unfortunately, you don't have enough blood to supply every tissue the maximal amount of blood it may need. Therefore, the body needs a way to divert blood to areas that need it the most. During exercise, for example, you need more blood sent to exercising muscles and less to places like your digestive system. How does your body accomplish this task? Well, it has some of its own "hose kinkers" in place; they simply need the right signals.

Noting blood flow control points

Blood vessels come in all sizes. Arteries and veins are the largest. Capillaries are the smallest; they're also the place where all oxygen extraction takes place. One intermediate-sized blood vessel helps control blood flow: the *meta-arterioles*. When you're thinking about blood flow control points, you need to be familiar with this blood vessel.

Meta-arterioles are the vessels that control the flow of blood by either squeezing down or by relaxing. They're smaller than arteries but larger than capillaries. Wrapped around the vessels are muscle bands, which squeeze tightly to reduce the amount of blood that can move through a vessel. They don't squeeze shut completely, but they can keep a lot of blood out (or let a lot in).

Factors that open and close blood vessels

The circulatory system is under constant influence from different stimuli to route blood flow to tissues of the body. Depending on which control system is activated, blood vessels may contract to divert blood flow, or they may relax to allow more blood to flow to the tissue. The primary controls are within the nervous system and chemical sensors in the blood vessels.

The nervous system's role in blood flow control

Nerves are necessary to shift blood. The two primary controls of the nervous system play a very important role in blood flow control. The parasympathetic nervous system (PNS) is most active at rest, and the sympathetic nervous system (SNS) is working during stress and exercise:

- ✔ The PNS relaxes blood vessels to areas of digestion, such as the gut, kidneys, pancreas, and even saliva glands. Blood flow to the skin is also reduced.

- ✔ The SNS increases blood flow to the skin (to get rid of heat), dilates arteries going to active muscle (to get more blood and oxygen to the tissue that needs it), and reduces blood flow to areas that are inactive.

These changes, which can happen quickly at the onset of exercise, help to provide more blood to the areas that need it without placing much additional strain on the cardiovascular system. (For more information on the PNS and SNS, refer to the earlier section "Fast or slow, what makes it go? The nervous system's influence on heart rate.")

Chemoreceptors: Sensors that know when you're working

Physical activity produces waste products like lactic acid and CO_2. A lack of oxygen to the tissue also triggers wastes due to anaerobic metabolism.

Have you ever had blood briefly shut off to an arm or leg? After you free the limb, you can feel the rush of blood because the arteries are wide open, waiting for blood. The waste products caused those blood vessels to open wide because of a response by chemoreceptors.

Chemoreceptors sense chemicals in the blood. These chemicals include the following:

- ✔ **pH:** The more acidic the area (due to a lot of lactic acid, for example), the more the blood vessels widen (called *vasodilation*).

- ✔ **Temperature:** The higher the temperature, the more vasodilation.

- ✔ **PCO_2:** Chemoreceptors sense the rise in carbon dioxide and cause dilation in order to bring more blood supply to the tissue.

Exercising and eating: Not a good combo

What happens when you eat something? Digestion! Digestion takes energy and requires blood to bring oxygen to the active tissue. (No surprise there. As we explain in this chapter, blood goes to areas of active tissue.) This is the reason you don't want to exercise on a full stomach.

If you begin exercising in the midst of digestion, not all blood is shifted away to the active muscle. Instead, the two active areas (your muscles and your gut) share the blood. As a result, your performance suffers. In addition, if the blood flow to your gut falls, you may be in for some trouble. Without proper digestion, the contents in your gastrointestinal tract will move along a bit too fast. Where's the bathroom!

One reason you want to warm up with some light activity before exercise is because doing so helps you activate the blood flow control mechanisms that deliver blood and oxygen to the muscles you'll use during the activity.

Noting the Effects of Exercise on the Cardiovascular System

The moment you start exercising, a multitude of changes begin to happen across your body as different systems come online to facilitate the activity. The cardiovascular system, as the engine of the work, will show rapid and immediate (also called *acute*) adjustments. The type of work being done influences the type of response. Aerobic and strength exercises provide two good examples to help you understand the changes that take place.

Acute adjustments to aerobic exercise

Take a boat out on a lake and push the accelerator. Off you go! Gas is burned, the engine roars, the prop turns, and the nose of the boat rises. After you get up to cruising speed, things level off: The motor stays at a steady roar, and the nose of the boat comes down to water level. Same thing with your body.

When you take off on a run or a ride, your heart kicks in and roars to life, pushing blood through the body, beating faster and stronger. Then after a few minutes, you seem to level off. These responses are the result of what's happening in your cardiovascular system. Take a look at the adjustments that occur.

As the levels of aerobic work increase

As the level of aerobic work increases, your body initiates a series of reactions that provide your cells with the oxygen they need to function under the increased workload. Here's what happens:

1. **The blood flow per minute, called *cardiac output*, increases.**

    Cardiac output is increased in a couple of ways — by increasing the heart rate and by increasing the stroke volume:

    - **Heart rate increases:** The PNS stops holding back heart rate, and the rate starts to rise. At the same time, the SNS kicks in fast, and heart rate rises even more. The stress hormone epinephrine is released, causing an additional rise in heart rate. The heart rate increases each time the workload increases until it can't go any higher, and you're at your limit!

    - **Stroke volume increases:** More blood fills the ventricle, so more blood can be pumped out. The SNS kicks in and causes the heart muscle to beat more forcefully, squeezing out more of the blood it has.

    Stroke volume has limits! Just as you can't squeeze a sponge dry, you can only squeeze so much blood out of a ventricle. Stroke volume hits its highest point usually around 50 percent of a person's peak aerobic fitness, or VO_2max. (Refer to Chapter 4 for more on aerobic fitness.)

2. **Your cardiac output rises and then stabilizes.**

    Because cardiac output is related to the amount of work you do, it stabilizes within a few minutes and matches blood flow per minute to the demands of the work. Remember that cardiac output is the product of heart rate multiplied by stroke volume, as we explain in the earlier section "Key measures of heart function."

3. **The active muscle gets more blood.**

    Blood vessels to active muscle dilate and bring more blood. The SNS, as well as the waste products of exercise (lactic acid, heat, and CO_2), are causing dilation.

4. **Your blood pressure (systolic) rises.**

    The added force of heart contraction creates a larger pressure wave during the contraction phase (systole). In a normal response, the systolic blood pressure rises with more work.

    Diastolic blood pressure, which is the pressure in between heart contractions, shouldn't change much. After all, nothing is different! Just the amount of time at rest changes. (The earlier section "Identifying the force behind the heart beat: Blood pressure" has more on systolic and diastolic pressures.)

5. Respiration (ventilation) rises.

Both the rate of breathing and the volume of each breath increase as work increases. More CO_2 is produced as a waste product, stimulating even more ventilation. Additional movement stimulates mechanical receptors, which also stimulate ventilation. If the work is heavy enough, lactic acid indirectly stimulates ventilation (because lactic acid can be converted to CO_2; refer to Chapter 4 for details).

6. Body temperature rises and sweating begins!

As the heat from energy production rises, blood flow to the skin increases. When body temperature exceeds an internal threshold, the hypothalamus in the brain, which senses body temperature, initiates a sweat response to dissipate the heat. The sweat comes from the water in the cells, the space between the cells, and the blood.

As the boat levels: The steady state

What happens if you just pick one work rate and stay at that work rate for, say, ten minutes? You may notice some adjusting going on the first few minutes, but things get better!

The point at which heart rate and oxygen consumption have caught up to the work demand and remain at a stable level is called the *steady state*. At steady state, heart rate, blood pressure, and most other variables remain stable (as long as you don't change the work rate).

Here's what happens to your heart rate, your oxygen consumption, and your breathing in the steady state:

- ✔ **Heart rate:** Your heart rate rises to meet the blood flow demand. This acceleration period often lasts a minute or two because that's generally how long the heart rate takes to catch up to the required work. After it does, your heart rate stays about the same for the rest of time you spend at that work rate.

- ✔ **VO_2:** Your oxygen consumption shows a similar rise at the beginning of work, but it, too, levels off after it catches up to the work demand.

- ✔ **Ventilation**: Your ventilation rises rapidly at the beginning of work, but it stabilizes after your system has matched the energy demand of the work.

If you're doing steady state work, cardiac output must stay the same. But in a hot environment, you sweat a lot! All that heat dissipation means that blood is pushed to the skin to help radiate heat, and blood plasma is lost as you continue to sweat. These changes result in a drop in stroke volume!

Fortunately, the body compensates for the drop in stroke volume by increasing heart rate to maintain a steady cardiac output. This upward rise in heart rate while at a steady state is called *cardiovascular drift*. An easy way to counter cardiovascular drift is to replace fluids to help offset the sweat loss.

Acute adjustments to strength training

Strength training is an anaerobic activity. Most efforts are quite intense and last only about a minute. During strength training, a number of things take place in the cardiovascular system, some of which may seem the opposite of what you may expect!

Strength training requires that many muscles contract. Contracted muscles squeeze down on blood vessels and produce a variety of effects. The magnitude of response depends on how much muscle is being contracted.

During the lift

The following steps outline what happens during the lift:

1. **The internal pressure slows the blood returning to the heart.**

   As the venous return drops, the heart has less blood to pump with each beat.

   Think of it like a big balloon of pressure in the middle of your body that holds blood in the limbs and briefly prevents it from returning to the heart.

2. **Blood is pushed out of the blood vessels and into the space between the cells, called the *interstitial space*.**

   Fluid being pushed into the interstitial space is a condition known as *edema*.

   Blood vessels can leak, and when they're under pressure, they can lose some of their plasma into the interstitial space. The plasma shifting to different locations actually results in reduced blood volume. You can often feel this effect. Ever wonder why your biceps feels tight and "pumped" after a lift? The trapped plasma in your interstitial space causes this sensation.

3. **Both systolic and diastolic pressures increase.**

   These can be quite high. Good thing this period is brief.

4. **Sensory nerves in the blood vessels sense the rise in pressure.**

   The receptors that sense the rise in pressure are called *baroreflexes*.

5. **Baroreflexes affect the heart and blood vessels in an attempt to counteract the high pressure.**

   Baroreflexes do the following:

   - They reduce heart rate (yes, heart rate goes down during the lift!)
   - They generate signals to the blood vessels to dilate.

No one wants high blood pressure, but during a weight lifting bout, it's hard to avoid, at least for a moment. However, you can inadvertently make matters worse by holding your breath. Holding your breath during a lift (called a *valsalva maneuver*) can greatly increase blood pressure, and it can be dangerous, especially for people with heart disease. Instead of holding your breath, breathe out in a smooth, controlled manner during the exertion portion of the lift. Breathe in on the lowering or the easier part of the lift. Doing so helps you maintain normal breathing patterns during exercise and helps minimize blood pressure increases.

After the lift

After the lift, the pressure drops rapidly, and more reflexes kick in:

1. **After the lift, when pressure drops quickly, the baroreflex kicks in again and does the opposite of what it did during the lift.**

   The baroreflex can sense both increases and decreases in blood pressure. After the lift, the following happens:

   - The drop in pressure causes an increase in heart rate. You may feel your heart rate take off just after setting the weight down.
   - Blood vessels starts to constrict to bring pressure back up. If pressure doesn't increase fast enough, you may even feel a little dizzy!

2. **Blood returns to the heart.**

   The pressure inside the body has been released, and all the blood held back in the limbs now rushes back to the heart.

3. **When the blood fills the right atrium, it causes a stretch, which initiates an reflex acceleration of heart rate (known as the *Bainbridge reflex*).**

   You may feel your heart is beating fast and strong, as it handles the excess blood that's returning to it.

4. **Edema sticks around, letting you feel pumped for awhile.**

   For about 30 minutes after the lift, you still have plasma trapped in the interstitial space. Your blood volume will remain a bit lower, at least until the fluid leaks back into the veins (via the lymphatic system).

You may realize now that, although weight lifting makes the heart rate increase, it is due to reflexes and pressure changes and not aerobic training. So if you want to do "cardio," don't rely on weight training. Do aerobic training!

Making Long-term Changes to Cardiovascular Performance

A wonderful thing about your cardiovascular system is that it adapts and grows stronger and more efficient if you give it a stimulus (like aerobic training). People who are inactive and out of shape can see huge improvements in their cardiovascular and aerobic abilities. Aerobic fitness comes about due to a number of specific adaptations, which we examine in the following sections.

To make changes in the cardiovascular and muscular systems, you need a well-balanced conditioning program that includes *both* aerobic and strength training.

Adapting to aerobic exercises

Exercise is a fantastic medicine for the body. By doing aerobic training, you can change many aspects of your cardiovascular system. It's like getting a complete overall to a car's engine! Just look at all the changes:

- **Heart rate:** Your resting heart rate is lower after aerobic training, because of the following factors:

    - **The PSN has become more dominant at rest.** Therefore, it slows the heart rate down more at rest than before. In addition, your heart rate comes back down to rest faster after a workout!

    - **The size of the heart chamber (ventricle) has grown.** With larger ventricles, fewer heart beats are needed to pump the same amount of blood at rest.

- **Blood pressure:** In people with high blood pressure, exercise can lower resting blood pressure. In some cases, it may lower blood pressure as much as 10 mmHg! Moderate aerobic exercise seems to exert the best effect.

- **Stroke volume:** Long-term aerobic training increases stroke volume (that is, more blood can be pumped with each stroke) because it

    - Helps to remodel the ventricles, enlarging the ventricles.

    - Strengthens the heart muscle, enabling it to pump out more blood.

- **Cardiac output:** Your maximal work output increases, partly due to the increase in stroke volume. If you can pump more blood, you can work harder!

Because cardiac output represents total blood flow and is related to the work you are doing, you only see improvements in cardiac output during work very hard. For more on cardiac output, refer to the earlier section "Key measures of heart function."

✔ **Blood vessels:** The density of capillaries in the muscles increases, meaning more oxygen can be delivered to the muscle. Just as making more water lines improves the irrigation of a farm field, aerobic training causes an increase in the irrigation of the muscle!

✔ **Oxygen extraction:** Aerobic training helps you use more of the oxygen your blood is bringing you. This adaptation is caused by

- The increase in the size and number of mitochondria, which draw the oxygen from the blood

- The increased availability of the oxygen (because you have so many blood vessels!)

Consistent aerobic training makes physical changes in the heart, in the blood vessels, and in your ability to use oxygen. All these changes happen at the same time and could never be replicated with a pill or supplement. Exercise is the best medicine for changing your body to become fit and to be able to do more work!

Adapting to strength training

Strength training is mostly an anaerobic activity and, therefore, doesn't really stress the cardiovascular system much. As a result, strength training doesn't produce many adaptations to the cardiovascular system. Heart rate, stroke volume, cardiac output, oxygen extraction . . . very little change is seen in these variables. Instead all the adaptations happen to the muscles, motor nerves, and the brain.

With strength training, your body adapts pretty specifically to the type of exercise you're doing. If you stress your muscles by strength training, then the skeletal muscles adapt. See Chapter 10 for more information on these adaptations.

Chapter 6

Earthlings and the Earth: Adapting to Your Environment

In This Chapter

▶ Understanding how the body adapts to hot and cold
▶ Looking at the effects on the body of exercising at altitude

*B*eing a warm-blooded mammal has its advantages. By generating your own internal body heat, you can maintain your metabolic function without having to rely on the external environment. Whether the day is cold or warm, your cells stay happy and functioning. But conditions can arise that make the body either too hot or too cold. Controlling body temperature can become quite difficult in extreme environments, a situation that can impair performance and even result in death.

In addition, as a creature that breathes air, you certainly value the oxygen that conveniently surrounds you, waiting to fill your lungs. However, as you ascend in altitude, the conditions change, and oxygen is not quite so available, a situation that presents problems for high-flying and high-climbing humans.

In this chapter, we explore how the body maintains its temperature under both hot and cold conditions and how it's affected by and acclimatizes to altitude changes. By the end of this chapter, you'll see how tough the environment can be on your body and the ways that your body can adapt to conditions to continue to perform its best.

Keeping It Just Right: The Basics of Temperature Regulation

Your body is finicky when it comes to temperature. It likes a very narrow range that hovers around 98.6 degrees Fahrenheit. Too far above or below that point, and cell function begins to decline, and systems don't function. Your body maintains an acceptable temperature through a series of control systems, starting with the hypothalamus in the brain. The hypothalamus signals other mechanism to either warm you up or cool you off.

The hypothalamus: Your internal thermostat

The hypothalamus is an almond-sized section of the brain located just above and slightly to the front of the brain stem. One of its key metabolic functions is body temperature regulation.

Blood continually courses through the hypothalamus, where its temperature is monitored. Each person has a hypothalamic thermostat set at a predetermined temperature, or *set point*. That set point is around 98.6 degrees Fahrenheit. The goal of the hypothalamus is to balance the gain of heat (due to exercise or the environment) with heat loss so that the set point is maintained.

Body temperature is not the same everywhere on the body. For example, the skin, also called the *shell*, is cooler than the blood, and the blood is cooler than active muscle and organs. The hypothalamus monitors the *core temperature*, the temperature deep within the body. The core temperature is the most important temperature to keep near its set point.

The hypothalamus does more than just regulate body temperature. It controls a range of functions like thirst, blood pressure, hunger, and hormone release. In some cases, the hypothalamus releases chemicals that impact the body directly (like releasing vasopressin to boost blood pressure), or it releases chemicals that cause other organs to release their hormones. For example, the hypothalamus may release a hormone that causes the pituitary gland to release growth hormone.

Pass the heat, please: The core-to-shell model of heat transfer

Hot things like to transfer their heat to cooler things. In order for the heat being generated in the muscle (lots of heat) to leave the body, it must come

Feeling feverish?

Although your set point is supposed to be maintained at 98.6°F, it can be modified. Under conditions of illness or infection, your body activates defenses to combat the nasty invaders. One of these self-defense tools is heat! The hypothalamus raises the body's set point, causing a rise in core temperature as high as 104°F. A high core temperature may fight the illness, but it can also be a danger to your cells if it goes too high — the reason why medications may be given to block the flames of a fever.

in contact with cooler and cooler locations. Therefore, the direction of heat movement looks something like this:

Heat in muscle → Cooler blood → Cooler skin → Cooler environment

As we explain in the next section, your body has ways of moving heat away from its core by cooling the shell (skin). Problems can arise, however, when these mechanisms fail or when extreme environmental conditions overwhelm the body.

Some Like It Hot — But Not Your Body!

Hot environments can pose a serious challenge to the body. If you add the heat generated during heavy exercise, you can be hit with a double whammy! Understanding the different ways your body gains heat from the environment can help you plan for the conditions.

Looking at the mechanisms of heat gain

The gain of heat by the body can come from within, or it can be from the external environment. Is the day sunny? Humid? Does the ground give off any heat? If the mechanisms of heat gain are combined, you could be in for a very hot day! This list shows the variety of ways you can gain heat:

- ✔ **Heavy exercise:** A large percentage — 80 percent, in fact — of the calories your body burns is lost as heat. So the harder you work, the more heat you produce.

- ✔ **Heat from the sun:** A sunny day can add 20 percent more heat load than a cloudy day. The sun heats the skin, making it harder to dispel the heat.

Heat dissipates by going from higher to lower temperature environments, as we explain in the earlier section "Pass the heat, please: The core-to-shell model of heat transfer." If your skin is warmer due to heat from the sun, the internal heat can't dissipate as efficiently.

✔ **Reflected heat:** Heat can reflect off many surfaces, like asphalt, metals, and water. (Ever run on a black asphalt surface? You may have noticed how hot it is!) The reflected heat is why stadiums can get so hot in the summer time, or why a city street is a lot hotter than a country lane.

✔ **Hot air:** A hot day can also warm the skin, making it hard for heat to get from the core to the skin. In this situation, the heat stays trapped inside the body (it's that core-to-shell model of heat transfer principle again).

✔ **Covered skin:** For heat to leave the skin, the skin must be exposed. Your clothing traps the hot air near your skin and prevents heat from leaving. Think about how hot it gets, even when you're wearing only a light shirt.

✔ **Humid days:** As the next section explains, one way to cool the skin is through sweating. The sweat itself doesn't cool you off; the sweat evaporating off the skin does the trick. Very humid air, however, already has all the water it wants, and it isn't willing to take your stinky sweat! On humid days, you can't cool your skin through evaporation.

Because the environment is such a contributor to heat gain, people try to measure the environmental conditions by using a variety of indexes. One is called the *heat index*. This index, shown in Figure 6-1, calculates a temperature by considering the air temperature and the relative humidity.

Temperature (°F)

Relative Humidity (%)	80	82	84	86	88	90	92	94	96	98	100	102	104	106	108	110
40	80	81	83	85	88	91	94	97	101	105	109	114	119	124	130	136
45	80	82	84	87	89	93	96	100	104	109	114	119	124	130	137	
50	81	83	85	88	91	95	99	103	108	113	118	124	131	137		
55	81	84	86	89	93	97	101	106	112	117	124	130	137			
60	82	84	88	91	95	100	105	110	116	123	129	137				
65	82	85	89	93	98	103	108	114	121	128	136					
70	83	86	90	95	100	105	112	119	128	134						
75	84	88	92	97	103	109	116	124	134							
80	84	89	94	100	106	113	121	129								
85	85	90	96	102	110	117	126	135								
90	86	91	98	105	113	122	131									
95	86	93	100	108	117	127										
100	87	95	103	112	121	132										

Figure 6-1: The heat index factors together air temperature and humidity.

Illustration courtesy of the National Weather Service

Turning on your personal air conditioner: The body's cooling mechanisms

To cool off, your body has a variety of mechanisms to move heat from your body's core to the shell. Some of these mechanisms work better than others:

- **Convection:** Hot air molecules can sit just above the skin, keeping the skin hot and preventing it from cooling. With *convection cooling*, air blows on the skin, sweeping away the hot air so that the skin can release more heat. If you've ever stood in front of a fan when you are really hot, you know what convective cooling feels like.

- **Conduction:** When something hot touches something cold, the heat moves to the colder object. In *conductive cooling*, the skin comes in physical contact with a cooler object — a cold towel, cold water, ice cubes, or anything else that can be touched — that transfers the heat away from the skin. Some people may even wear cooling vests, which can go underneath clothing. Ever wonder how firefighters stay cool? Cooling vests!

- **Radiation:** Infrared rays can warm a fast-food hamburger, and they also can take heat away from the skin. Remember, your blood is warmer than your skin. When the core temperature rises, the hypothalamus signals blood vessels in the skin to open wider, which lets more hot blood move to the skin. When the hot blood gets near the surface, its heat can radiate away. Another effect of this cooling mechanism is that your skin looks red; that flushed appearance is the result of the widened blood vessels. Hopefully, you'll never get warm enough to cook a burger!

- **Evaporation of sweat:** Water vapor holds far more heat than water itself (think of steam, which is hotter even than the boiling water that produces it). When your core temperature rises above the set point, the hypothalamus triggers sweat glands in the skin to release fluid (sweat) onto the skin. This fluid has been heated by the blood coursing through the skin. When it reaches the surface of the skin, it turns to a vapor (evaporates) and takes the heat with it.

Sweat evaporation, which does 80 percent of the work in cooling the body, is the most important cooling mechanism your body has. The more skin you have exposed to the environment, the more sweat evaporation can take place (remember how humidity affects evaporation, though).

Adding insult to injury: Exercising in hot environments

Athletes can't hide from the heat. They often must perform in conditions that would make most of us run inside for some iced tea. If their bodies' cooling mechanisms are working, things may go just fine. However, if the athletes are unaccustomed to the heat or if they can't cool themselves, a rise in body temperature above the set point, or *hyperthermia,* is inevitable, a situation that results in a decline in performance and possible serious injury.

When athletes encounter a hot environment for the first time, they notice an immediate effect on performance. Heat reduces their ability to perform well.

Seeing the effects of a hot environment on the body

Exercising in a hot environment produces certain changes in the body:

- ✔ **A faster rate of glycogen (carbohydrate) use:** As carbs are used more quickly, athletes deplete the primary energy source for much of their work. Have you ever heard of the marathon runner that "bonks," "crashes," or "hits the wall" with fatigue during a race in hot weather? He has run out of glycogen! (Chapter 4 has more information on glycogen and exercise metabolism.)

- ✔ **Reduced muscle blood flow:** With so much blood going to the skin to help cool the body, less blood is available for the muscles. Less blood flowing to the muscles means less oxygen for the muscles, fewer nutrients, and reduced performance.

- ✔ **Increased blood lactic acid:** Because the muscle is starving for oxygen, it relies on anaerobic metabolism to keep going. The resulting lactic acid contributes to bonking and a very unpleasant feeling in the muscles. (For a detailed explanation of what lactic acid is and how it affects the body during exercise, refer to Chapter 4.)

- ✔ **High rate of water and electrolyte loss through sweating:** Sweat evaporation is the body's primary cooling mechanism. In the untrained individual, the sweat contains not only water but also electrolytes (sodium and potassium), which are important for cell function and the transport of nerve signals. They must be maintained within the body.

Dehydration reduces performance. Every volume of sweat lost results in a reduction in performance. You also run the risk of heat injury. To prevent dehydration, you must replace fluids lost as quickly as possible, both during and after the activity.

- ✔ **Increased heart rate:** With blood going to the skin for cooling and more fluid being lost due to sweating, a reduction in blood volume occurs. With less blood volume, the heart must beat even faster than usual to push the same amount of blood.

Too hot to handle: Heat injury

Athletes are known to push themselves very hard, sometimes ignoring warning signs of injury. When this happens in a hot environment, the result can be very bad. Even hours after being in the heat, the illness can continue.

Determining how much fluid you need to replace

Did you know that one fluid liter weighs one kilogram? Additionally, one pound of fluid is 16 ounces. So to determine how much fluid is lost due to sweating, you do the following:

1. **Weigh the person (with minimal clothing) before the activity.**

2. **Weigh the person after the activity.**

3. **Subtract the after-exercising value from the before-exercising value.**

   The difference in weight represents the loss of fluid!

Here's an example: Before a long run, Bob weighs 180 pounds wearing just his running shorts. After the run, he weighs 177 pounds:

180 pounds − 177 pounds = 3 pounds fluid lost = 48 ounces that need to be replaced (3 × 16 = 48).

Get Bob drinking! Remember that he has also lost things like sodium and potassium, so sports drinks may be a nice solution to the problem.

Heat injury is a leading cause of death among athletes, and the injury has three primary stages of severity, listed here from least to most severe. Action should be taken at the first signs of heat injury:

✔ **Stage 1 — Heat cramps and dizziness:** During the early stages of heat injury, the cells become dehydrated, and fluid is lost from the blood. The result is a feeling of nausea and dizziness, as well as muscle cramps.

Treatment: Get the person out of the hot environment! Give her cool fluids as soon as possible, massage the cramps, and have her stretch. If she is dizzy, have her lie down.

✔ **Stage 2 — Heat exhaustion:** If the athlete pushes past cramps and ignores the nausea and dizziness, things can become quite severe. Heat exhaustion causes extreme losses in fluid. Symptoms of heat exhaustion include the following:

   • Cold clammy skin, despite a high internal temperature

   • Pale complexion

   • Collapsing from dizziness or even losing consciousness

   • Headache and nausea

   • Weak and rapid pulse

   • Shallow breathing

Treatment: Call 911 immediately. Heat exhaustion is very dangerous and needs immediate care by a physician. The person with heat exhaustion may need so much fluid that it must be administered intravenously. The sufferer's sweating mechanism is shutting down due to the loss of so much fluid, and the condition can progress to heat stroke (and death)

quickly, so you must act fast! Until emergency responders arrive, cool the sufferer with cold water and, if available, ice.

✔ **Stage 3 — Heatstroke:** At this point, the person has collapsed completely and her ability to cool herself has shut down. Symptoms of heat stroke include

- Very high body temperature (106°F or more)

- Dry skin (sweating has shut down)

- Unresponsiveness

- Labored breathing

Treatment: Without immediate treatment, death can occur quickly (often due to things like liver or kidney failure). Call 911 immediately, and, if possible, immerse the sufferer in ice. Your goal is to cool her down fast!

Getting your body to adapt to the heat

Fortunately, your body is equipped to adapt to hot environments. It only takes about 10 to 12 days, and the recipe is pretty straightforward:

✔ **You must exercise in the hot environment.** Sorry, sitting in a sauna or working on your tan does not count! Adaptations only occur when you work in the actual environment.

✔ **Take it easy and work up intensity.** Whatever the training intensity you are used to, back it off to about 60 percent to 70 percent for three to five days. As your body starts to adapt, you can build up intensity.

✔ **Show some skin.** Wear minimal clothing to help your skin cool through sweat evaporation. Clothing should be light, breathable, and reflect light. If you normally wear pads (you play football, for example), start without them for a while.

✔ **Take water breaks.** Stay hydrated! Drinking about 8.5 ounces of water every 15 minutes helps offset sweat losses.

Within 10 to 12 days, your body will acclimatize. The changes that occur, listed here, help cool your body faster and allow better (and safer) performance:

✔ **Sweat rate increases.** Your sweat glands put out more sweat, which means more cooling.

✔ **Sweating begins earlier.** The sweat mechanism kicks in more quickly, so you don't get too hot before things start to cool.

✔ **Sweat is more watery.** The sweat glands reduce the amount of sodium and other substances in the sweat, a combo that helps conserve electrolytes. Watery sweat also evaporates faster, which is cool — literally and metaphorically!

✔ **Plasma volume increases.** The kidneys retain more fluid for the blood, and because some of the sweat comes from the blood, this translates into more fluid volume.

✔ **Your body uses less glycogen.** Glycogen use drops after acclimatization, which means you don't fatigue as fast. (Glycogen, which is a stored form of glucose in the muscle and liver, is discussed in greater detail in Chapter 4.)

✔ **Your skin is cooler.** Due to all the preceding changes, the skin stays cooler, enabling heat to move from the core to the shell more easily.

✔ **Your heart rate is reduced.** Because of all the fluid increases and improved cooling, the heart doesn't have to work as hard, and your performance in the hot environment will be similar to your performance in a cooler environment.

When Chillin' Ain't Cool: Exercising in Cold Environments

Humans are capable of surviving in space, so it's not hard to believe that cold environments *can* be conquered. Obviously, standing in Antarctica in your running shorts isn't a wise move, and certainly exercising in cold environments has risks that you must be aware of. But proper clothing and understanding how much heat you generate during the activity can go a long way toward ensuring that you safely handle cold environments.

Baby, it's cold inside: Introducing hypothermia and wind chill

When heat loss exceeds the body's ability to generate heat, a condition of hypothermia exists. In *hypothermia,* the body temperature drops below 95°F, and a loss of body function results. Causes of hypothermia include

✔ Exposing skin to cold air, which rapidly draws heat away

✔ Exposing skin to cold water, which dramatically accelerates heat loss

✔ Wearing wet, cold clothing, or falling into cold water

Being both wet *and* cold rapidly lowers body temperature. If you cannot generate heat to offset the drop, you risk hypothermia! Symptoms of hypothermia include the following:

- ✔ **Shivering:** The muscle are being activated to rapidly contract as a way to generate heat.

- ✔ **Fatigue and sleepiness:** These symptoms indicate that the nervous system is beginning to slow down, as are body functions.

- ✔ **Slowing heart rate and breathing:** The central nervous system is slowing its function. If body temperature drops too much, breathing can stop entirely — and so can your ticker!

- ✔ **Unconsciousness:** Let's hope you are getting help by this point!

Hypothermia progresses through three stages:

- ✔ **Stage 1 — Mild:** This stage may almost seem like a stress response as the body tries to offset the drop in core temperature. It involves shivering, fast heart rate, and an increased use of glucose.

- ✔ **Stage 2 — Moderate:** Body temperature keeps dropping. The shivering becomes pretty severe as the body forces muscle contractions to generate heat. Coordination of movement becomes slow and error prone, and people in this stage may experience mild confusion. You may notice that the sufferer looks pale and has very cold hands. The reason is that the body is constricting blood vessels to the body's shell as a way to keep the core warm. Lips may even look pale or blue.

- ✔ **Stage 3 — Severe:** The body begins to shut down. Heart rate drops very low, respiration falls, and blood pressure drops. Confusion, disorientation, uncoordinated muscles, and weakness mean that the person is probably not in a position to save him- or herself. Organ failure and death can follow.

Treat hypothermia before it becomes a problem. If you see someone who is likely suffering from hypothermia, get him dry, wrap him up, and give him heat! Your goal is to warm his core (his chest and trunk) and to get him away from the frigid environment!

Wind chill can become a problem when the combination of air temperature and wind is such that skin freezes quickly if exposed to the elements. You may feel all warm on the inside, but wonder why your fingers are frozen. When air temperature gets down around 10°F, it takes only a small breeze to create conditions that can freeze the skin within minutes. Ultimately, how severe the problem is depends on air temperature, how warm you are, the length of exposure, and the wind chill.

Keep the heat: Dressing for the cold

The key for success in cold weather exercise is insulation. You have to cover the skin and provide layers of insulation to keep the heat. Here are some suggestions:

- ✔ **Layer clothing.** Air is an insulator. By layering clothing, you create layers of air, which help hold heat in.

- ✔ **Wear clothing that can wick away moisture.** Wet clothing loses its air pockets of insulation. Newer clothing materials let water escape without losing the air insulation layer. It's worth the extra money for some nicer winter materials, especially if you exercise outdoors often or frequently engage in outdoor, cold-weather activities!

- ✔ **Adjust clothing based on how much heat will be produced during the activity.** A heavy work, high-heat producing activity requires less insulation than a light work, low-heat producing activity. Remember, dressing too heavy could actually bring on hyperthermia (refer to the earlier section "Adding insult to injury: Exercising in hot environments" for details on this condition)!

Kids lose heat much more quickly than adults do, so keep them well insulated. Older folks also may get colder faster and not generate as much heat, so be sure to wrap them well!

A happy lung is a warm, moist one!

Try breathing on a window. See the moisture? Your lungs like to be warm and wet. But cold air sucks the moisture right out of them. Dry and irritated lungs get stiff and narrow and make breathing hard. The result may be *cold air induced asthma*, a condition that can make breathing very difficult, reduce exercise capacity, and create uncomfortable situations for just about anyone.

So how can you avoid cold air induced asthma? One simple treatment is to put a scarf or mask over your mouth while you exercise. Doing so warms the air slightly and keeps more moisture from escaping. Drinking extra water in cold climates can also offset the water lost in the breath.

Live High and Train Low: Exercising at Different Altitudes

Humans are a finicky lot. We can't live in the ocean, don't like the desert much, and are mostly confined to warm climates under 10,000 feet. So when we do venture outside our comfort zone, the body's ability to do work is immediately impacted. Fortunately, we can adapt (sort of), which helps extend our range of wandering this planet.

Oxygen is the currency needed for life. Normally, people live within an ocean of air (oxygen, carbon dioxide, and nitrogen). But as you travel higher into the atmosphere, oxygen has a harder time getting into your blood and muscles. With less oxygen to help you do work, your performance suffers (or worse). The following sections have the details.

Revisiting oxygen transport

As we explain in detail in Chapter 5, pressure is the key behind all gas movement. Air always flows from high pressure to low pressure. The highest pressure of oxygen is in the atmosphere; next is your lungs; and then the blood. Oxygen is "pushed" from your lungs into your blood based on a pressure difference, and from there, it's pushed into your muscle cells:

Oxygen in atmosphere → Air in lungs → Blood → Muscle cells

Because humans aren't fish and can't take oxygen from a fluid, we need something to carry the oxygen. That carrier is *hemoglobin*. Hemoglobin can pick up a bubble of oxygen and transport it to the muscle cells. It contains iron, which bonds to the oxygen and gives you nice, red blood. All the hemoglobin needs to do its job is a nice push (pressure). Refer to Chapter 5 for details on the role hemoglobin plays in transporting oxygen.

As the pressure of oxygen in the atmosphere drops — as it does at higher altitudes — less oxygen is pushed into the blood, resulting in less oxygen for the hemoglobin to carry.

Each gas that makes up air (oxygen, carbon dioxide, and nitrogen) exerts a pressure independent of other gases. The individual pressure for each gas is known as that gas's *partial pressure*. Here are the partial pressure values of each of these three gases (for details on how to calculate partial pressure, refer to Chapter 5):

- ✔ **Oxygen:** 159 mmHg
- ✔ **Carbon dioxide:** 0.03 mmHg
- ✔ **Nitrogen:** 600.7 mmHg

Barometric pressure is the pressure exerted by the total of all the gases, and it's 760 mmHg at sea level.

Partial pressure can be modified, either by changing the concentration of the gas (like adding a bottle of oxygen) or by changing the pressure. So when you are exercising or climbing at altitude (which drops partial pressure), you can use some supplemental oxygen from a breathing system and increase the partial pressure. Doing so allows mountain climbers to actually climb higher than typically possible — provided they don't run out of their extra oxygen!

When going up brings you down: Altitude and reduced aerobic capacity

Problems caused by low air pressure at high altitudes don't really begin until you get above 5,000 feet, but once above that altitude, not enough pressure exists to push oxygen fully into the blood. As a result, less oxygen is available to you for exercise. As you go higher, you experience more pronounced effects:

- ✔ **At 10,000 feet:** Your fitness is about 20 percent lower than at sea level. You run slower and get tired faster. Even a simple walk takes more effort.

- ✔ **At 14,000 feet:** Your aerobic ability is 30 percent weaker. If you are out climbing in Colorado, just hiking up a hill may feel like you're running your hardest. You have to stop periodically just to catch your breath. Your heart rate may be near its max.

- ✔ **At or above 20,000 feet:** At this level, few people function very well unless they are acclimatized. Simply not enough oxygen is being pushed into the blood to enable you to do much work.

Why am I alive in a jet flying at 40,000 feet?

Jets fly at very high altitudes and take air into the cabin at that altitude, and people manage just fine, even though a person would normally die at 40,000 feet. What gives? Pressure!

The concentration of oxygen in the air at 40,000 feet is the same as at sea level, but there is so little pressure that the oxygen can't get pushed into the blood. To counter this situation, jets have cabins that are pressurized to the equivalent of an altitude of about 8,000 feet — enough pressure that most people don't have any problems. However, doctors may suggest that people with heart and lung ailments avoid flying for fear of issues that may arise due to the slightly less-than-ideal conditions.

You can avoid the drop in atmospheric pressure as you go up in altitude. Head to the later section "I think I can, I think I can . . . Adapting to high altitudes" for information. Still, due to the loss of oxygen and reduced ability to do work, training at high altitude doesn't make much sense. You won't be working very hard and, therefore, can't improve your training!

A sick view from the top: Identifying altitude illnesses

Not everyone has difficulty adjusting to changes in altitude, but that doesn't mean that, when you and the family go skiing in Colorado at 10,000 feet, someone isn't at risk for some of the following symptoms of altitude sickness:

- **Dehydration:** Dehydration is common for two reasons. One, the air is dry higher up, and you lose water simply by breathing. Two, your kidneys release fluid (you hit the bathroom more) as a way to concentrate the blood to carry more oxygen.

- **Headaches:** Mild headaches are common with dehydration. If these get worse and progress to significant pain and debilitation, watch out. You could be sensitive to more serious conditions.

- **Fatigue and insomnia:** These symptoms usually go away within two days as you adjust.

When arriving at higher altitude, take a few days to adjust, drink lots of water, and don't exert yourself too much.

In the following sections, we outline some of the more serious conditions that can occur at high altitudes.

High-altitude pulmonary edema

At high altitudes, some people are susceptible to more serious conditions that are life threatening. These individuals experience changes in their lung tissue that cause fluid to accumulate in the space between the lungs and the blood, blocking oxygen from getting to the blood. Called *high-altitude pulmonary edema,* the condition produces these symptoms:

- Coughing and flu-like symptoms

- Difficulty breathing (you gasp for breath)

- Skin (lips and fingers) that may look blue

- A tight chest and wheezing, which indicate the lungs have fluid surrounding them

If these symptoms occur, the single best treatment is to get down from the altitude! Get below 3,000 feet.

High-altitude cerebral edema

Whereas high-altitude pulmonary edema is an accumulation of fluid in the lungs, *high-altitude cerebral edema* is the accumulation of fluid in the brain. This condition is often fatal, so take it seriously! If you see these symptoms, get down from the altitude immediately:

- ✔ Nausea and/or vomiting
- ✔ Weakness, confusion, and/or loss of memory
- ✔ Severe headache
- ✔ Balance problems
- ✔ Hallucinations

The preceding symptoms are all signs that your brain function is in trouble. The only treatment that works? *Get down from the altitude!*

I think I can, I think I can . . . Adapting to high altitudes

Given time at high altitudes, your body will acclimatize. How much time depends on just how high you are going. Altitude acclimatization takes longer the higher up you are. Hiking at 10,000 feet, for example, may mean waiting three to four days for acclimatization to occur. When you're climbing at 20,000 feet, you may need a month or more before your body adjusts.

So what changes occur during acclimatization to high altitudes? The key is oxygen. You can't change the atmosphere, so the only way to acclimatize to altitude is to improve your body's ability to load up on oxygen and deliver it to the cells. Key adaptations to altitude include the following:

- ✔ **New blood vessels form.** Yes, your body actually produces new blood vessels to help deliver more oxygen to the muscles so that they don't starve.

- ✔ **Red blood cell production increases.** More red blood cells means more hemoglobin, which in turn means you can carry more oxygen. Even though the push of oxygen is less, you make up for it by having more hemoglobin carrying oxygen. Many hands make light work, the saying goes. This is your body's best adaptation to high altitudes.

- ✔ **Blood vessels dilate.** Wider blood vessels can deliver more oxygen to the tissue and help keep you moving.

The drug that did more than help altitude illness

Researchers in the early 1990s were working on a medication that could dilate blood vessels in the lungs and heart to help reduce fluid around the lung as well as deliver more oxygen to the heart. Clinical studies started to show an interesting side effect. Patients reported that they had erectile improvement as a side effect. Research on the altitude illness treatment was abandoned, and a drug, later called Viagra, was developed. This story goes to show that new discoveries come via unusual pathways!

Living high and training low: The best of both worlds

So you want to be the best you can be? Well, then, you need to take advantage of the best that sea level and altitude can give you. We call this the *Live High, Train Low* method, and it lets you use the enhanced oxygen delivery adaptations that occur at high altitudes while you exercise at low altitudes:

- ✔ **Live high:** Achieving the adaptations that enable your blood to carry more oxygen takes time living in an altitude environment. Some people even sleep in tents that have reduced pressure to simulate altitude (that's getting pretty serious).

- ✔ **Train low:** Your strongest training happens only when you have the most oxygen available, so you train at lower altitudes. By taking advantage of the adaptations in your blood to carry more oxygen while you train at lower altitudes, you can push your fitness and performance to the max!

This method gives you the best of both worlds. It may not be for everyone (we are *not* suggesting you commute back and forth between the mountains and sea level!). However, some top aerobic athletes do try to take advantage of the Live High, Train Low method.

Part III
Basic Biomechanics: Why You Move the Way You Do

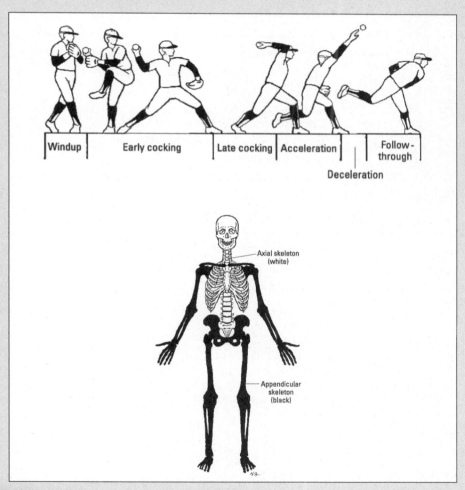

Windup | Early cocking | Late cocking | Acceleration | Follow-through

Deceleration

Axial skeleton (white)

Appendicular skeleton (black)

In this part...

- Distinguish between the fields of kinetics and kinematics
- Identify the kinds of motions that exist and distinguish between types of joints and their functions
- Discover how muscles are oriented, how they're recruited to contract, and what they're made of
- Uncover why muscles fatigue and how they can get bigger and stronger
- Analyze movement and functional activities (like making a jump shot), using a systematic approach

Chapter 7

The Nuts and Bolts of Movement

The world is continually moving. Every day you can look around and notice people doing things like walking, running, cycling, and swimming. You probably haven't given much thought to what it takes to perform these skills. Few people actually sit and analyze how to walk, run, cycle, or swim, let alone how to do any of these activities better — unless they've been forced to through an injury that immobilized them or they are in a field where performing at peak ability is vital.

Movement and the factors that affect motion are quite complex and require an understanding of several principles related to physics. For example, everything that you encounter is being acted on by forces. Regardless of how something may be moving (or not) — whether it's spinning, flying through the air, or sitting perfectly still — invisible forces are affecting it. This chapter provides insights into what is behind getting stronger, throwing a ball, and running faster, among other things. You'll never again think the same about how you move!

Biomechanics: The Study of Movement

You've no doubt heard of Newton's laws, forces, and vectors in high school science or physics courses. When you learned about these principles, chances are you did so in relation to nonhuman examples. Well, the field of biomechanics takes those principles and applies them to human movement.

Biomechanics is the investigation of how forces act on the human body from a mechanical perspective. The *mechanics* part of the term represents the study of physical actions, and the *bio* part refers to living organisms (in this case, humans). Examining the effects of forces, their impacts on the human body, and how the human body creates motion either internally or externally is a fascinating and very dynamic area of investigation.

By studying biomechanics, you can answer questions like, "Why do some baseball pitchers hurt their elbows?" "How can I make myself run faster (or jump higher)?" "Why don't I fall off my bike when I lean really far over as I go around a corner?" and "How come the Olympic gymnasts can do such amazing acrobatic things and look so graceful?" — a question I (Brian) ask myself repeatedly as my daughter tries to teach me how to do a cartwheel.

The role of the biomechanist

We all appreciate the grace and power that Olympians and professional athletes exhibit. But these athletes aren't as good as they are because they were born that way. Sure, genetics has something to do with their abilities, but they've also studied their skills and investigated how to improve. In a sense, they have been their own biomechanists.

Shooting a basketball, walking, jumping, and skipping all sound like very simple tasks, but they are all actually quite complex skills. Executing any task involves the interplay among multiple forces and body positioning. How far you should bend your elbow or knees, how big your steps should be, and which foot goes first are the kinds of things you start to ponder when you learn these skills.

By understanding the types of forces and motion in relation to the structure and function of the human body, you can look for ways to enhance performance and prevent injury.

The biomechanist's problem-solving process

To assess any type of activity or task, biomechanists follow a process. Typically, this process starts with understanding the nature of the task, followed by a deliberate observation of the task in action. From this observation, data is collected and used to evaluate the performance, which ultimately leads to feedback and, if necessary, an intervention. The next sections take you through this process.

Understanding the nature and objective of the task

The first requirement in assessing a movement is to understand the task being completed. You must have this prerequisite information to fully address what, in fact, is occurring as part of the movement. For example, if you want to jump, you know that you first need to bend at your knees, which causes your ankles and hips to also bend and puts you in a sort of squatting position. From this position, you then push up as hard as you can, a movement that makes your body rise and, if you have enough force behind the push, lifts you off the ground.

When you know how the joints move during the activity, you can start asking more complex questions like, "How do I jump higher?" This type of question requires not only an understanding of joint motions but also knowledge about which muscles are involved and how they're activated.

In addition to having an understanding of the task being performed, the biomechanist must understand the intent of the activity. For example, analyzing a pitcher's throwing motion can only be done with an appreciation of what the pitcher is trying to accomplish.

Observing the task and collecting data

Biomechanists must be able to collect the necessary data. Necessary data is any data that provide substantive and defendable information that is key to answering the question.

Suppose a pitcher comes into the athletic training room because of a sore elbow. The athletic trainer must collect the data necessary to answer why the pitcher's elbow hurts. Key information includes things like how often the pitcher throws, what type of pitches he throws, and whether he was throwing long toss or off the pitching mound. The answers to these kinds of questions help the athletic trainer fully understand the situation being assessed.

Evaluating the data and making a diagnosis

After observing the task and collecting data, biomechanists use the data to make comparisons of the current situation to others like it. Everyone walks or runs differently, but certain things have consistently been proven to affect the success of the task. In this step, biomechanists compare the specific data they gathered from a subject to what they know should be occurring. Based on this information, they can recognize flaws and identify areas of improvement.

The assessment is a critical aspect to answering your question. When you perform such an assessment, you must do so objectively, taking the information (data) as it is and using your findings to answer the question.

Sharing the findings with the athlete: Intervention and feedback

Concluding the process, biomechanists share their findings with their patients/athletes. By sharing the information soon after the activity and in a way that identifies the key components, a plan then can be developed to improve performance and, in many cases, avoid injury.

Kinematics: A Compass Telling You Where You Are

To assess and evaluate motion, you must understand the underlying components of human structure and the various physical and environmental components that affect everyone. *Kinematics* is the study of movement and includes considerations of form, velocity, direction, time, acceleration, pattern, displacement, and sequencing.

Many refer to this area of study as the descriptive, or qualitative, portion of motion analysis because values aren't quantified. Kinematic references are made in relation to things like direction and how fast someone may be going, or what movement is happening first and what follows. In kinematics, you don't consider the forces within the movement, which makes kinematics different from kinetics, a topic that we introduce in the later section "Studying Kinetics: May the Force Be with You!"

Looking at body systems

Activities of daily living, like brushing your teeth, rising from a chair, or pouring a glass of water, are necessary to normal, everyday living. Although they are often taken for granted, especially by healthy, active people, each task is actually very complex. In fact, any movement you make, even the "ordinary" ones, requires coordination and communication between your body's muscular, skeletal, and nervous systems:

✔ **The muscular system:** This system involves the soft tissue structures that are *contractile* in nature; that is, they have the ability to create tension by contracting. Muscles are connected to bones by tendons. When a muscle is contracted (see Chapter 11), it contributes to or produces an end result: Often muscle contraction results in joint movement. Other times, muscle contraction results in no motion but helps to support the body and increases stability.

Assessment in action: How do I run faster?

Figuring out how to run faster isn't as easy as you may think. Using the process outlined in the section "The biomechanist's problem-solving process," you would approach this problem in this way:

1. **Understand the nature and objective of the activity and amass the necessary prerequisite information.**

   To more deeply understand the act of running, you need to ask yourself what actually occurs during this activity. Clearly, you have to move your ankles, knees, and hips. But what other components are involved? Breaking down the running (gait) pattern can help. What about the type of running — is your client a distance runner or a sprinter?

2. **Observe the task and collect data.**

   Typically, the gait cycle is broken down into two phases: stance and swing (refer to Chapter 3). The stance phase begins with your foot striking the ground and continues as your body weight is transferred forward. It ends when you push off at your toes to propel yourself. While one foot is in the stance phase, the other is in the swing phase.

   The gait cycle specific to running varies from that of walking in that the heel strike doesn't exist. Instead, striking occurs at the forefoot (by the toes). Additionally, the weight transition phase is either nonexistent or occurs in a very small window of time. The interesting aspect of running is that, at times, the body is totally airborne, and no contact with the ground occurs — a situation that doesn't happen when walking.

Understanding the motions that make up the gait cycle is important, but how those motions are achieved is equally valuable information that can help you answer the question, "How do I run faster?" Muscle actions during the gait cycle dictate the amount of push-off during propulsion at the ankle and knee. Additionally, the muscles bring a limb from the swing to the stance phase, prepare for landing, and help maintain balance throughout.

By understanding the motion and the muscle forces that are needed to complete this task, you can begin to address the question. You now know that running speed is dependent on the length of each of your strides, the force exerted during the push-off coupled with the resistance (body weight), and how fast you can bring your leg back after you push off.

3. **Evaluate the data and make a diagnosis.**

   Noticing that someone may be taking strides that are entirely too long or are really bouncy gives you information that you can use to identify flaws. By comparing the current task to what you know about how the task *should* be carried out, you can diagnosis areas for improvement.

4. **Provide intervention and feedback.**

   Given the information that you gathered and the results of the evaluation and diagnosis, you can now provide feedback and suggest an intervention program. This feedback and intervention may be as simple as telling the client to shorten his stride when running or to not bounce so high when he goes from step to step.

✔ **The skeletal system:** The skeletal system provides the framework that supports the human body. Your bones are the structural blocks that allow you to stand, sit, and walk. Your muscles attach to these bones, and when they contract, the bones are pulled by the tendons, resulting in movement of the joint.

✔ **The nervous system:** The nervous system, often overlooked and largely taken for granted, involves the brain and subsequent nerve function throughout the body. Absolutely everything that you do is controlled by your nervous system. Your brain initiates movement by deciding the strategy that is needed for the particular motion, and the peripheral nervous system acts as the highway to transmit the signals among your brain, spinal cord, and the rest of your body. The brain helps determine motor patterns (for information on motor control, head to Chapter 3), and this system interprets pain, activates muscles, and controls motion.

Each of these systems has very important individual functions, but it's their coordinated work that enables you to go about everyday as you do. Consider something as simple as walking. The motor patterns that dictate walking are defined in the brain from past activities and are sent to the local areas (knee, ankle, and hip), where the movement is carried out. Here's what happens:

1. **As soon as you decide that you want to walk, your brain interprets your body position, where you are (walking up a hill, stairs, and so on), and decides how to best move.**

2. **Each of your muscles is activated to propel your body. While the motor pattern is being sent to the muscles that are active, other muscles are preparing for their role.**

   At the same time you may be pushing (propelling) yourself forward from your toes, for example, your other foot is preparing for your heel to strike the ground and your same-side hamstring group is readying to slow your leg down.

3. **All the while, your skeletal system serves as the source of support and conduit for movement.**

Identifying forms of motion

When analyzing motion, biomechanists often divide movement into three subcategories: linear motion, angular motion, and general motion. Figure 7-1 shows the different types of motion:

Linear motion

In *linear motion,* all parts of an object move in the same direction at the same velocity (speed and direction). Often this type of motion is referred to as *translation.* Linear motion can occur in two ways, rectilinear and curvilinear:

✔ **Rectilinear motion** occurs when the object in its entirety is moving in a straight line. A passenger sitting still on a train that is going straight is a good example of moving in a rectilinear fashion.

✔ **Curvilinear motion** results in a uniformly curved pattern of movement. Curvilinear motion exists when a stunt driver, for example, takes his car and jumps over a flaming bonfire. Because of the momentum generated while going up the ramp, the car goes up and over the fire and lands on the other side. As long as the car doesn't spin as it goes over the fire, the motion is curvilinear.

Rectilinear

Curvilinear

Angular

Figure 7-1:
Types of motion: rectilinear, curvilinear, angular, and general.

General

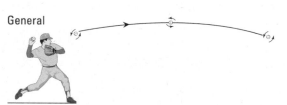

Illustration by Wiley, Composition Services Graphics

Angular motion

Moving objects don't all travel in a uniform direction. Instead, they often involve angular motion as well. Angular motion is the movement of an object around an axis of rotation or an imaginary line.

Nearly every motion that occurs within the body is an example of angular motion. When muscles pull on the bones, they cause the bones or limb segments to bend or rotate at their joints. A figure skater performing a spin on the ice is an example of angular motion. The skater's entire body is spinning in relation to the axis of rotation while she balances on the ice.

General motion

A majority of the motion that occurs everyday falls into the general motion category. *General motion* exists when linear and angular motions are combined. A ball thrown by a pitcher is an example of both angular and curvilinear motion: The spin of the ball is angular motion, while its trajectory as it approaches the batter is curvilinear (an arc).

Human motion is almost always general motion. As you walk down the street, your joints experience angular motion (because they're swinging in an arc) while your body as a whole is moving in a straight line *(translating)* down the street in rectilinear fashion.

Defining key terms

To understand and break down the movements that exist in the world, biomechanists must have command of the vocabulary. When referring to areas of the body or movements, you do so in reference to their positioning, called *anatomical position*. Anatomical position is an upright standing position that has the feet separated shoulder width apart, the arms hanging at the sides, and the palms facing forward (see Figure 7-2).

Using directional terminology

To describe how the body moves and the relationship of an object to the body, you must be able to use the following terminology:

- **Superior:** Closer to the head, or "above." Your shoulder is superior to your hip.

- **Inferior:** Closer to the feet or "below." Your knee is inferior to your hip.

- **Anterior:** Toward the front of the body. Your nose is anterior to the back of your head.

- **Posterior:** Toward the back of the body. Your heel is posterior to your big toe.

- **Medial:** Toward the middle of the body. Your nose is medial to your ear.

- **Lateral:** Away from the middle of the body. Your ear is lateral to your mouth.

- **Proximal:** Closer to your trunk. Your elbow is proximal to your wrist.

✔ **Distal:** Further from your trunk. Your foot is distal to your knee.

✔ **Superficial:** Closer to the surface of the body. Your sternum is superficial to your heart.

✔ **Deep:** Away from the surface of the body. Your intestines are deep to your abdominal muscles.

✔ **Axial:** Along the longitudinal axis of the body. An acorn falling onto the top of your head is an axially directed force.

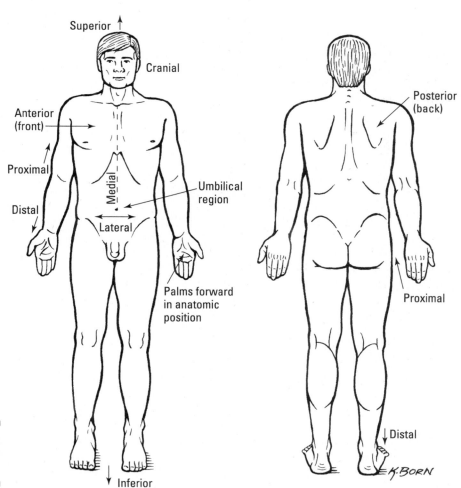

Figure 7-2:
Anatomical
position.

Illustration by Kathryn Born, MA

Planes of motion

When standing in anatomical position, the body is bisected by three cardinal reference planes, shown in Figure 7-3. A *plane* is an imaginary flat surface that divides the body into equal halves. The three planes (sagittal, transverse, and frontal) all intersect at the point considered as the center of mass. The planes are used as a reference in relation to the human body. When the body moves, the planes also move with it. The planes are

- ✔ **The sagittal plane:** This plane travels down the body and divides it into left and right halves.

- ✔ **The transverse plane:** This plane separates the body into top and bottom portions.

- ✔ **The frontal plane:** This plane travels down and separates the body into front and back portions.

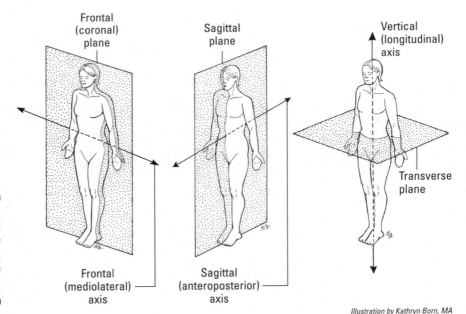

Figure 7-3: Cardinal planes of motion and axes of rotation.

Frontal (coronal) plane

Sagittal plane

Vertical (longitudinal) axis

Transverse plane

Frontal (mediolateral) axis

Sagittal (anteroposterior) axis

Illustration by Kathryn Born, MA

References to body motion are often based on the plane within which the motion occurs. Motion that moves in-line with a particular plane or parallel to it is typically referred to as motion occurring in that plane.

To determine in which plane the motion occurs, imagine that the motion was completed standing next to mirrors representing the planes of motion. The mirror(s) *not* broken after the movement would be considered the plane that the motion occurred in.

Jumping jacks, for example, are a frontal plane movement. During a jumping jack, your arms and legs move outward to the side from the middle of your body (*abducted* — we get to that term in a minute). If you do a jumping jack in front of mirrors representing the planes of motion, you would break the sagittal and transverse plane mirrors as your arms and legs move out to the side, but your frontal plane would be untouched.

Axes of rotation

When you move your limbs (arms and legs), you do so around an imaginary axis of rotation that passes through the joint. Three axes of rotation are used in describing motion; each of the axes is orientated perpendicular to one of the three planes of motion (refer to Figure 7-3):

- ✔ **Anteroposterior axis:** Rotation in the anteroposterior axis is perpendicular to the frontal plane.

- ✔ **Mediolateral axis:** Rotation in the mediolateral axis occurs perpendicular to the sagittal plane.

- ✔ **Vertical axis:** Rotation in the vertical axis occurs perpendicular to the transverse plane.

The axis of rotation is perpendicular to the plane of motion because this arrangement allows for the limb to rotate around a hinge or axle. If you take a look at a tire on a vehicle, you note that an axel runs from left to right under the car, and on each end is a tire that's orientated perpendicular to the axel. This arrangement allows the car to move forward when the tire rotates around the axel. Your joints all bend and create rotation, and the rotation happens around an axis (like a door hinge or axle). You define an axis of rotation by the direction in which the axis runs.

To explain in what axis of rotation a specific motion is occurring, consider the orientation of the movement and how the movement can be generated about the axis of rotation. For example, moving your elbow closer to your upper arm *(flexion)* is a sagittal plane movement and requires an axis that runs from side to side through the elbow. That axis would be the mediolateral axis.

Joint motions

Believe it or not, each movement that you make can be defined very precisely. If someone were to ask you to bend your elbow, you'd have no trouble understanding what that means, but if you think about the term *bend,* you realize it's actually pretty imprecise — too imprecise, in fact, to be used as a descriptor when you're analyzing motion. This particular motion is actually called *flexion,* a term that has a very specific meaning in motion analysis.

Joint motion definitions allow you to specify what is actually happening within human movement. The term *flexion,* for example, means more than just "to bend." It also describes that the bones that make up the joint are moving closer to one another. Without the terminology to define aspects of motion to this level of specificity, biomechanists would not be able to explain types of motion in a way that guarantees understanding.

Biomechanists consider many motions, based on which cardinal plane the motion is executed in. The following sections list the most common movements.

Movements in the sagittal plane

Movements that occur in the sagittal plane are typically described as joints moving closer together in an anterior or posterior direction. Following are the most common sagittal plane motions, which you can see in Figure 7-4:

- **Flexion:** *Flexion* consists of a movement within the mediolateral axes of rotation that brings the joint together or lessens the joint angle.

  The term *flexion* also describes two separate motions at the ankle. *Dorsiflexion* is the act of raising your toes toward the sky, and *plantar flexion* is the act of pointing out your toes. These toe motions go directly against all the rules of describing motion, so you just need to memorize them separately.

- **Extension:** *Extension* is a movement in which a greater angle is created at the joint.

- **Hyperextension:** Hyperextension is a movement that makes the joint move past its anatomical position. Often, people describe this kind of motion as "double-jointed."

Movements in the frontal plane

Movements in the frontal plane are those that either move away or toward the midline of the body (see Figure 7-5). The most common of frontal plane movements are *abduction:* and *adduction*

- **Abduction:** When you move your limb away from the midline of the body, or in a lateral direction

- **Adduction:** When you move the limb toward the midline of the body, or in a medial direction

The hip and shoulder are great examples of places where abduction and adduction movements are common. The spine (lateral flexion), scapula (elevation/depression), wrist (ulnar deviation and radial deviation), and ankle (eversion/inversion) also contribute to frontal plane movement.

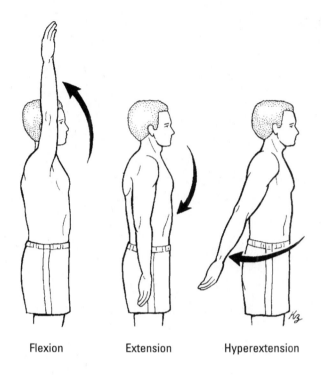

Flexion   Extension   Hyperextension

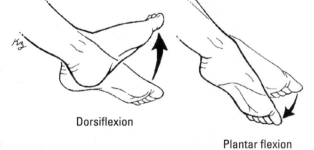

Figure 7-4:
Sagittal
plane
movements.

Dorsiflexion

Plantar flexion

Illustration by Kathryn Born, MA

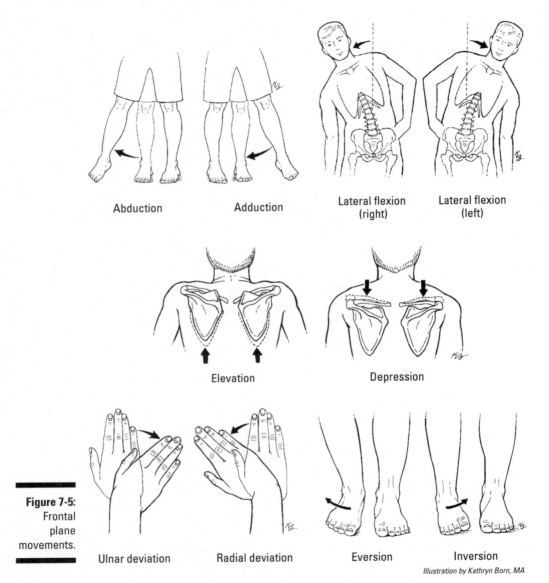

Abduction   Adduction   Lateral flexion   Lateral flexion
                                         (right)   (left)

Elevation   Depression

Figure 7-5:
Frontal
plane
movements.

Ulnar deviation   Radial deviation   Eversion   Inversion

Illustration by Kathryn Born, MA

Movements in the transverse plane

Rotation around the vertical access is said to exist within the transverse plane. Rotation also occurs in the spine (right/left), shoulder (internal/external), hip (internal/external), and the wrist (supination/pronation).

Internal rotation is often also referred to as *medial rotation,* and external is often called *lateral rotation,* as shown in Figure 7-6.

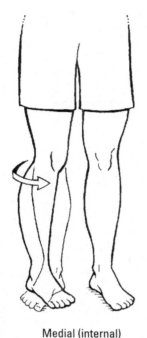 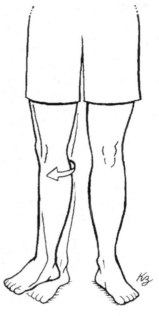

Figure 7-6:
Traverse
plane
movements.

Medial (internal)
rotation

Lateral (external)
rotation

Other kinds of movement

No dynamic or otherwise functional task occurs solely in one of the three cardinal planes. Instead, a great deal of motion is diagonal. *Diagonal motion* happens when a movement occurs in more than one plane. For example, throwing a baseball is a movement that crosses the body in a diagonal fashion.

Circumduction is one of those "other" movements that don't fit within one particular plane but relate across each of them. Circumduction occurs when the limb is directed in an imaginary circle, not as in spinning around (like a top), but as in the whole segment tracing a circular pattern with the distal aspect. Imagine pinwheeling your arms. That's circumduction.

Newton's toolkit: Lever systems

Lever systems are everywhere — in the world and in your body. A lever system is a rigid segment that rotates about an axis. In this context, the axis of rotation is referred to as a *fulcrum*. A lever system relies on both a force being applied (called *effort)* and an opposing force *(resistance)* that you intend to move.

When you use a pry bar, for example, you apply force when you pull down on the bar, and the resistance is the boulder itself. The fulcrum is the point that's in contact with the ground, where the pry bar rotates when you pull down.

Just like the pry bar serving as a lever, your body also consists of numerous levers and lever systems that assist you in all types of activities. Within the body, your bones serve as the lever, the joints are the axis of rotation (or fulcrum), and your body (or what you are holding) serves as the resistance.

Types of levers

How the three parts — force, resistance, and fulcrum — are applied to a lever defines the type of lever and action that it creates (see Figure 7-7):

✔ **First-class lever:** In a first-class lever, the fulcrum is placed between the force and the resistance. A common example is the teeter-totter.

An example of a first-class lever in your body is your head sitting on top of your spine. The spine serves as the fulcrum, the posterior neck muscles are the force, and the front of the head serves as the resistance.

✔ **Second-class lever:** In a second-class lever, the resistance is between the fulcrum and force. You use this type of lever to move heavy resistance with relatively small force. A common, everyday example of a second-class lever is a wheelbarrow. The wheel is the fulcrum, the resistance is what's in the wheelbarrow itself, and the force is what you exert on the handles to lift and move the wheelbarrow.

Within the body, the most common second-class lever is your foot and ankle. If you rise up on your toes, the ball of your foot, where it bends, is the fulcrum. Meanwhile the force (resistance) of the body is applied to the proximal foot area (from the tibia), and the force (effort) allowing you to go up on your toes is coming from the attachment site of the calf muscles to the heel.

✔ **Third-class lever:** In a third-class lever, the most common lever system within the body, the force is placed between the fulcrum and resistance. This type of lever allows for the greatest speed and range of motion during the task. It's also one of the trickier ones to understand because of where the force is applied. In a third-class lever, the force needed to create the movement is actually in contact with the lever.

A good example of a third-class lever is the elbow flexion that occurs during a dumbbell curl. Your elbow is the fulcrum, and the resistance is the thing you're holding onto, like a dumbbell. In this case, the elbow flexors (biceps brachii) exert force (effort) onto the lower arm, just distal to the elbow joint. Because of this arrangement, the elbow can both move quickly and far, based on its lever system.

First-class levers

Second-class levers

Third-class levers

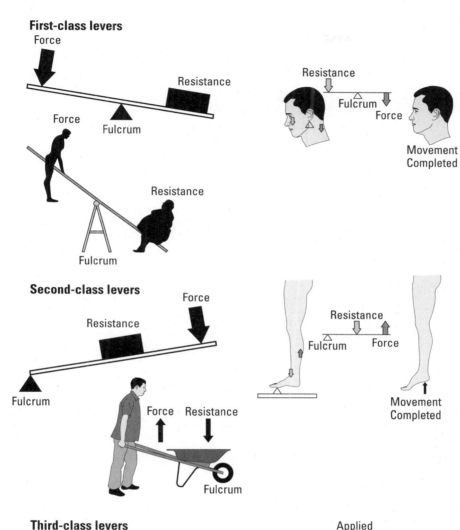

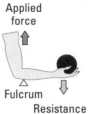

Figure 7-7: Types of levers.

Illustration by Wiley, Composition Services Graphics

Manipulating levers for maximum advantage

Components within each lever system can be manipulated to allow for more advantage. When the force and resistance are manipulated in a way that allows the task to be completed more easily, you've achieved a mechanical advantage.

Force arm and resistance arm dynamics

The amount of force (effort) needed to move the resistance is determined by the distance its application is from the fulcrum and the weight of the object. The *force arm (FA)* is the distance from the point of *force application* (the point where the force is applied to the lever) and the axis of rotation. The *resistance arm (RA)* is the distance from the point of resistance to the axis of rotation.

The force needed to move an object is inversely proportional to the length of the FA. The longer the FA, the less force needed to move the object. The longer the RA, the more force needed to move it.

Your body must be quite strong to do many of the daily tasks that you need it to, yet the human body contains many third-class levers where the RA is longer than the FA. This clearly puts you at an instant mechanical disadvantage. For example, the FA at the elbow is dictated by the biceps brachii muscles' attachment to the radius, which is only 2 inches or so away from the elbow joint (fulcrum). Because the insertion is so close to the fulcrum, the elbow's function relies on a great deal of strength. Short of surgically lengthening the FA in your joints, you can't do much about it besides strengthening yourself.

A majority of the sporting-related tasks involves several different types of levers acting at the same time to complete the task. For example, when you jump, the levers at the toe, ankle, knee, and hip joints must each work to accomplish their own unique goals while simultaneously *not* working against the other joints.

Velocity

Velocity refers to the speed and direction of a body or object. The longer the force arm, the more effective it is in imparting velocity. A higher velocity results in a higher force being imparted on the object at contact. For example, a baseball player can hit the ball a greater distance when he uses a longer bat instead of a shorter one. But there's a tradeoff: Although the longer lever imparts a higher velocity, it also increases the time needed to complete the task. (For more on velocity, head to the later section "Velocity and acceleration.")

When a thrower needs to make a faster throw to get a runner out, the thrower shortens his lever arm length (he bends his elbow) and snaps the ball out quickly. By shortening the lever, the player is able to complete the throw in a shorter time.

Are you really double-jointed?

You've probably seen someone who can contort his or her body into what seem to be superhuman ways; maybe you can even do this yourself. People, like the circus performer who folds herself into a box or the classmate who bends his fingers or arms in ways that stun or disgust you, are said to be "double-jointed." In fact, no one is double-jointed.

So how do people put themselves in these awkward positions and look so out of joint? The ability to move in such ways is usually dictated by flexibility. *Flexibility* refers to how inherently extensible the soft tissue of the body is. Your joints are held together by ligaments and other soft tissues. These tissues can vary in the amount of support they provide. The person who claims to be double-jointed is simply very flexible and probably has very lax joints.

Although shortening the force arm helps you perform a task more quickly, it isn't the right strategy to use when you need to produce a high amount of force, like making a long throw from the outfield to home plate.

Balance, equilibrium, and stability

To affect movement, levers are dependent on the force and resistance applied to the fulcrum. Yet the goal isn't always to move an object. Sometimes, the goal is to allow an object to remain in its current state. Keeping objects in their current state requires balance, equilibrium, and stability:

- ✔ **Equilibrium:** The state of zero acceleration, where no change occurs in the speed or direction of the body

- ✔ **Balance:** The ability to control equilibrium, either *statically* (while still) or *dynamically* (while moving)

- ✔ **Stability:** The resistance to a change in the body's acceleration

To understand the relationship between these three factors, think about a boat being rocked back and forth by the strong ocean winds and waves. Despite being rocked back and forth, the boat must retain its balance (its original upright position). The shape of the hull and the sail allow the boat to sway side to side but maintain stability to avoid tipping over; the weight of the boat itself also provides stability by resisting the effect of the waves.

These three concepts are very important to maintaining good joint health. Your ligaments, in particular, keep the joint functioning in the direction that it is meant to. For instance, the medial collateral ligament, which runs on the inside of the knee, keeps the knee from bending inward. If stability is compromised in this ligament, the knee becomes injured.

Feeling displaced and distant

In kinematics, you need to be familiar with the concepts of displacement and distance. *Distance* is the total length traveled by an object from point A to point B. *Displacement* is the length from point A to point B. Examining distance and displacement is key when you want to evaluate a task or motion for efficiency.

To understand the difference between distance and displacement, think about what happens when a punt returner catches the football: He runs forward, changing direction a number of times to avoid tacklers. From beginning (catching the ball) to end (getting tackled), he may run a total of 30 yards because of all the ducking and weaving, but he may move only 5 yards forward. The 30 yards he traveled is his distance, and the 5 yards moving forward is his displacement.

Measuring kinematics

Kinematic analysis examines the *qualitative* (descriptive) and *quantitative* (numerical value) aspects of the movement. Doing this type of analysis lets you evaluate limb movements, see how fast they may be moving, identify in what sequence they occur, and so on.

To gather this information, biomechanists use a high-speed video camera or digital movement capture system, which allows them to capture more frames per second than the standard video camera. The high-speed capturing systems can provide more information and allow for clearer assessments to be made.

Anyone can pull out a camera these days and record someone running, batting, or otherwise completing an activity. From the video, you are able to assess the range of motion of the joints, the sequencing of the activity, and things like limb and body orientation. True, you won't have the details that are available to professional biomechanists, but you can still get a good idea of what's going on.

Tasks like walking require specific amounts of motion to avoid injury and maximize performance. Another form of kinematic analysis happens in the rehabilitation clinic and after surgery in your doctor's office when they measure your joint range of motion. They compare this range with the normative values known for each joint.

Studying Kinetics: May the Force Be with You!

Kinetics is the study of the impact that different kinds of forces have on mechanical systems — in this context, your body. Beyond seeing the ball spinning, kinetics offers insight in to the how and why the ball is spinning.

The secrets of movement

Behind every great movement are the principles and components that affect it. Concepts like inertia, mass, force, weight, pressure, volume, density, torque, impulse, displacement, distance, and center of gravity are the backbone of movement.

Inertia

Inertia is the concept that states that objects tend to want to stay in their current state of motion, whether they're still or moving, until another force acts on them. A bowling ball sitting still on the floor, for example, will stay that way until another force (a push by a bowler) acts on it. The same holds true for a football flying through the air. It will remain in motion unless other forces, like gravity, air friction, or a wide receiver, make it stop.

 Inertia is an object's resistance to change. The amount of inertia an object possesses is directly proportional to its mass. The heavier an object is, the more force required to change it from its current state.

Mass

Mass is the amount of matter contained in the object. Not all objects of the same size have the same amount of mass or the same weight. A ping pong ball, for example, is about the same size as a golf ball, and the golf ball weighs more. So in this context, the golf ball possesses more mass, making it heavier.

Force

The heavier an object is, the more force is required to move it. *Force* is the push or pull that acts on a body or object; the formula for force says that its product equals that of its mass and its acceleration:

force = mass × acceleration

Things like gravity, wind, and water are all forces that can act on you or other objects.

Center of gravity

The *center of gravity* is the area within an object or body that is positioned at the middle of the object's mass. The center of gravity exists regardless of an object's position and is the point by which you measure motion and the effects of any applied external forces. When biomechanists investigate motion in a gymnast, for example, they concentrate on the gymnast's center of gravity, not on her arms and legs.

Weight

How much an object weighs is determined by the amount of gravitational force being applied to it. (This is why you weigh less on the moon, which has less gravitational pull, than on Earth, which has more.) *Weight* is considered a body's relative mass or the quantity of matter contained by it.

Weight is often confused with mass. The key difference is that weight is a force, and mass is simply the amount of matter within an object. Weight is always oriented toward the center of the earth because it is directly related to gravity.

Torque

Levers are rigid systems that turn about an axis. To turn, levers rely on torque. *Torque,* the turning effect of force, is dependent on the application of an *eccentric force,* a force that is applied to an object but is not oriented with the center of rotation of the axis (see Figure 7-8). The more torque applied to the axis of rotation, the greater the turning effect of the force.

Force directed downward leads to resultant rotation.

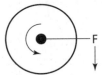

Upward directed force of the biceps on the arm results in joint rotation

Figure 7-8: Torque is dependent on an eccentric force.

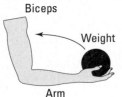

Illustration by Wiley, Composition Services Graphics

Impulse

Forces can be applied to objects in varying amounts of time. The *impulse* is the amount of force applied during a unit of time. A large force may have a relatively insignificant effect on an object if it's delivered only for a split second. Conversely, a fairly small force may have a great effect on an object if it's delivered continuously over time.

Suppose, for example, that you made the mistake of kicking a medicine ball (a large, heavy, solid ball) really hard. You probably experienced a great deal of pain in your knee because the ball didn't move. If you had used a lesser force and applied it over time, you would have discovered that moving the ball across the weight room and out of your way was relatively easy — and pain free.

Velocity and acceleration

Forces are often *translated,* or transferred, to objects. To fully understand the effects of force application, you need to explore the delivery of the force, its speed, and the direction in which the force is applied.

Velocity is the quantity that represents the speed and direction of a body or object. A concept directly related to velocity is that of acceleration. *Acceleration* is the rate at which an object changes velocity.

Every movement has a speed component; many are slow, and some are very, very fast. This speed in relation to movement's direction and how it may change can have significant influence on the resulting force and its effects on an object.

What a load!

Different types of forces can be delivered to the body. The effects of the forces on the body are referred to as *mechanical loads.* By understanding the types of forces and their effects on the body and its structures, you can understand regular function. Figure 7-9 shows a variety of mechanical loads.

✔ **Compression:** *Compression* refers to an *axial force* (a force directed along the longitudinal axis of the object) that results in squeezing or pressing.

Did you know that in the morning you are typically taller than you are at the end of the day? You can thank compression for that. During the day, you're upright, and gravity and your body weight compress you all day long. You actually shrink over the course of a day!

✔ **Tension:** The pulling of a force on an object is *tension.* The muscles often apply tension as they pull on bones to make them move. In fact, the many bumps *(tuberosities)* on your bones are a result of muscles pulling on the bone.

✔ **Shear:** A force directed perpendicular to the longitudinal axis of an object is a *shear force.* The result of a shear force allows one portion of the object to displace in relation to other portions of the object. Shear forces often result in bone fractures. If, for example, a force is delivered to the shin when your foot is fixed to the ground, one portion of the shin may be displaced in relation to the other.

✔ **Bending:** *Bending* involves the combination of tension and compression. When a non-axial force is applied to an object, bending occurs. In other words, when one side of an object experiences compression while the other side undergoes tension, it bends.

✔ **Torsion:** When one end of an object is fixed and it experiences twisting at its longitudinal axis, *torsion* occurs. Injuries that involve torsion are pretty common in sports, especially when the playing surface has a lot of traction.

✔ **Combined loads:** The human body experiences a number of different types of loads (force application) simultaneously during many activities. *Combined loading* occurs when more than one type of load is delivered to an object. Combined loading is the most common in the body.

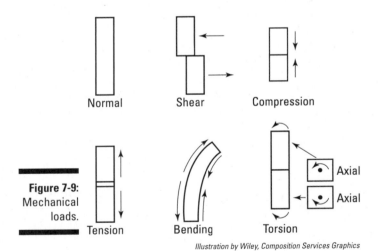

Figure 7-9: Mechanical loads.

Illustration by Wiley, Composition Services Graphics

Newton's laws of motion

Throughout this chapter, we explain forces and the effects they have on objects, including the human body. In a nutshell, inanimate objects move because forces are applied to them; the body moves in the same manner. Although human bodies don't commonly roll around or fly through the air like balls do, they do move because forces — both internal and external — are applied:

✔ **Internal forces are the forces generated within the body.** For example, when muscles contract, they administer a pulling force to the bones that results in movement.

✔ **External forces are force originating from outside the body.** For example, when you fall, you do so because you've lost your balance, and gravity pushes you down.

Force application and the subsequent movement or lack thereof is dependent on Newton's three laws of motion, which govern the principles of all types of motion.

Newton's first law: The law of inertia

The first law of motion states that an object at rest or in uniform motion (no acceleration) will remain at rest until acted on by an outside force. Because this law represents an object's resistance to changes in its current state (motion or still), this law is commonly referred to as the law of inertia.

Think about moving a heavy object across the garage floor. You probably notice that getting the object moving takes a lot more force than keeping it moving. The higher force needed to first move the object is an example of inertia and the object's resistance to the movement.

Newton's second law: The law of acceleration

The second law of motion states that the acceleration of an object is directly proportional to the force acting on it but inversely proportionate to the mass of the body. What this means is that moving a lighter object is easier than moving a heavier one. If you've ever had to move to a new residence, you can easily appreciate this law. I (Brian), for example, always seem to end up with the boxes that have more mass in them and that require the exertion of more force.

Newton's third law: The law of action and reaction

Newton's third law states that for every action there is an equal and opposite reaction. Translation? Whenever an object exerts a force (action) onto another object, the contacted object exerts a reaction force that is in the opposite direction and of the same magnitude.

Think about an arm wrestling contest. If both competitors are of the same strength and therefore exert the same amount of force, then the net movement is zero. However, as the force exerted by one of the competitors lessens (because this competitor fatigues and weakens), the force exerted by the opponent increases, and he wins.

Sometimes the reaction isn't as obvious, because we tend not to think of inanimate objects "exerting" anything: So thinking about what happens when you walk or jump may help you grasp the concept. During either of these tasks, you push down against the ground to propel yourself forward. When you push down, the ground "pushes" back against you with the same force, which enables you to move forward or jump up into the air.

Measuring kinetics

Just as the types and actions of forces are varied, so are the ways to measure them. Depending on the question, biomechanists determine the type of variable (the specific force) they need to measure and how to go about doing it. The methodology or process a scientist follows often determines whether the information collected will actually answer the question being asked. Part of that methodology is choosing how to measure the particular force or its effects.

Many biomechanists, for example, examine muscle activity when they want to know what exercise leads to the most amount of muscle firing. Or they may be interested in the pattern of your walk (your gait), in which case they use a force plate platform for high speed video or motion analysis equipment.

Most of the ways to measure things within the realm of kinetics are very expensive and require very sophisticated equipment, not to mention someone who is highly trained in using the equipment.

Ain't it a kick in the head? Concussions

You can't go anywhere today without hearing about concussions. The National Football League has recently instituted many new rule changes both in the game and in how medical professionals treat patients with concussions. Additionally, many states have introduced legislation that dictates how and by whom patients can be released back into participation after sustaining a concussion.

Several researchers are dedicated to investigating concussions and understanding how this injury should best be identified and treated. A recent trend in this research is investigating the actual magnitude of the blow delivered to the heads of football players. Participants in these studies are given helmets outfitted with force transducers or accelerometers that are capable of measuring the acceleration of the head and its direction. By collecting this data, researchers can examine the forces and subsequent injury these athletes may sustain.

Some commercial companies are now marketing this type of technology to be widely available and utilized in a number of different ways. For instance, one ad shows such a helmet being used by a young boy skiing downhill. When the boy sustains a large enough force to his helmet, a message is sent to his parents' smartphones, alerting them that the boy has been injured and advising them that he should be referred to a physician for follow up. Using the kinetic information available through such devices can help save lives and decrease the long-lasting effects of concussion injury.

Chapter 8

Bone Composition and Function

· ·

In This Chapter

▶ Types of bones and their function

▶ Materials that make up bones

▶ How your bones grow

· ·

A h, the human skeletal system! It's made up of 206 bones that perform two major functions within the body: The bones support and protect the body and the organs within it, and they provide rigid attachments for muscles and other soft tissues that, altogether, create and support movement. Beyond these primary functions, your bones dictate other, very important aspects of your life. For example, the bones play a critical role in *hemopoiesis,* or the formation of blood cells. And as they model and remodel constantly to accommodate the demands your body places on them, your bones contribute significantly to your daily life.

In this chapter, we discuss the types of bones that exist, identify what they are made of, and explain the means by which they grow and accommodate your activities.

Boning Up on the Basics

Most people think of bones as hard, unforgiving structures. What they often overlook is just how dynamic and accommodating bones really are. Much of a bone's structure is defined and dictated by its function. As you look closely at a picture of a skeleton, you can see that some bones are intended to assist with load bearing (like walking and running), and others aren't.

You can generally recognize whether a bone is load bearing or not by its size and shape. Load-bearing bones are typically bigger and thicker, whereas non-load–bearing bones are usually smaller. But the outside appearance of the bone isn't the only indication of its function; the inside also provides clues.

Collectively, bones help deliver nutrients, store minerals, protect your organs, support your posture, and make stability and movement possible. A bone's life is dynamic; it constantly adjusts to its environment and the tasks that you demand of it. *Wolff's law,* named after German anatomist Julius Wolff (1836–1902), describes the dynamic adaptation of bone. It states that a bone will adapt to the loads that it encounters — as long as the loads themselves are not so large as to break the bone, of course!

Looking at bone's composition

The foundations of most bones are calcium carbonate, calcium phosphate, collagen, and water. The percentage of each of these substances vary by the function, age, and relative health of the person and the bone itself. Here's what you need to know:

- **Calcium carbonate and calcium phosphate** are the minerals primarily responsible for the bone's strength, and they account for approximately 60 percent to 70 percent of the bone's weight. They help dictate its *stiffness* (the extent to which it resists deformation from a force) and *compressive strength* (its ability to resist being squeezed or otherwise shortened), both of which are important determinants of a bone's function. Although these materials play a primary role in a bone's makeup, other materials such as sodium, magnesium, and fluoride also help.

- **Collagen** (a protein) is responsible for the bone's flexibility and ability to resist *tension* (a pulling force).

- **Water** enhances the strength of the materials that make up the bone and contributes significantly to nutrient delivery and waste removal. Water accounts for about 25 percent to 30 percent of a healthy bone's weight. When the water content decreases, as it does with aging, bones become more brittle.

Think about what happens when you try to break a limb that's been recently trimmed from a tree. It doesn't break cleanly; you have to bend it repeatedly back and forth to get it to break. But wait a few days to break the limb, and you can do so easily. The reason is the lack of water. Just as a tree branch becomes easier to break as it dries, bones becomes more brittle and easier to break as they age.

Pouring over porosity: Cortical versus trabecular bones

Bones are characterized based on their amount of porosity. In the context of bone, *porosity* refers to the number of cavities or pores that exist in the bone, and it determines the bone's strength. A bone that is porous possesses less calcium carbonate and calcium phosphate.

Level of porosity also determines whether a bone is classified as a cortical bone or trabecular bone:

- **Cortical bone:** A cortical bone, which has a low porosity, is very dense and mineralized; typically only 5 percent to 30 percent of its volume is non-mineralized. A cortical bone is much stiffer than a bone with high porosity and is able to resist greater *stress* (pressure or force exerted on the bone).

  Consider the femur, the largest long bone in the body. It's responsible for absorbing the great deal of force generated when you walk or run; therefore, a large part of its makeup is cortical bone. The femur can absorb a significant amount of pressure and tension (stress), but it can't be bent very far before it breaks (strain).

- **Trabecular bone:** A trabecular bone has a higher percentage of non-mineralized tissue. Also referred to *cancellous* or *spongy,* trabecular bone has a very porous makeup. Bone marrow exists in this type of the bone. In addition, because of its decreased levels of mineralization, trabecular bone can absorb more *strain* (the measure of deformation in the bone) than cortical bone before it breaks.

No one bone is exclusively cortical or trabecular. Instead, throughout each bone, various levels of porosity exist. Cortical bone typically resides in the outer shaft of the bone, and the spongy trabecular portion usually exists further within the bone. The location and amount of porosity found in the bone is determined by the amount and type of forces that the bone absorbs. Because their structural organization is a response to the various forces exerted on them, bones are considered *anisotropic*. Bottom line: Bone is a dynamic structure that constantly adapts to the forces applied to it.

The structural components of bone

The following list outlines the structural components of bones (see Figure 8-1):

- **Articular surfaces:** The *articular surfaces* of long bones, the parts that form the joints, are located at the proximal and distal end of the bones and are typically made up of a higher percentage of trabecular bone.

- **Diaphysis:** The long bone's shaft, called the *diaphysis,* consists of an outer shell (the *cortex*) that is covered by a dense fibrous connective tissue called *periosteum*.

- **Periosteum:** The *periosteum,* which covers the diaphysis, is very dense cortical bone that is highly sensitized by *nociceptors* (sensory neurons responsible for the perception of pain). That's why breaking a bone hurts so much.

- **Cortex:** The medullary cavity houses bone marrow. The cells lining the medullary cavity — the periosteum and endosteum — are collectively called the *cortex*. The density of the cortex is high because this area of the bone needs to be so strong.

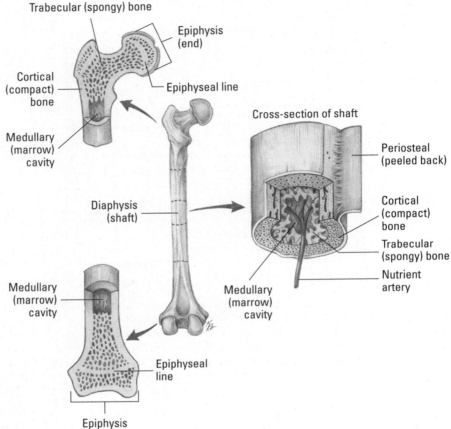

Trabecular (spongy) bone

Epiphysis (end)

Epiphyseal line

Cortical (compact) bone

Medullary (marrow) cavity

Diaphysis (shaft)

Medullary (marrow) cavity

Cross-section of shaft

Periosteal (peeled back)

Cortical (compact) bone

Trabecular (spongy) bone

Nutrient artery

Medullary (marrow) cavity

Epiphyseal line

Medullary (marrow) cavity

Epiphysis (end)

Figure 8-1: Long bone structure.

Illustration by Kathryn Born, MA

- ✔ **Metaphysis:** Marrow is a flexible, spongy tissue that's responsible for creating red blood cells through a process called *hematopoiesis*. A majority of hematopoietic activity occurs in the *metaphysis,* or the ends, of the long bone.

- ✔ **Epiphyseal plate:** At the point where the diaphysis and metaphysis join is a line of cartilage that separates the sections. This line, referred to as the *epiphyseal plate* (or, more commonly, the *growth plate)* is the site of longitudinal bone growth (or the lengthening of the bone).

- ✔ **Apophysis:** Where tendons and/or muscles insert onto the bone, a raised section, called an *apophysis* exists. An apophysis is separated from the cortex by an apophyseal plate that consists of *osteoblasts,* bone growth cells. The most prominent example of an apophysis is the *tibial tuberosity* — the slightly raised section on the bottom of your knee that you can feel just by running your hand over your knee and down your shin. The tension that the quadriceps puts on this area by pulling on the tibia via the patellar tendon causes new bone cells to be laid down, forming the raised portion.

You're stressing me out! Compression, tension, torsion, and shearing

Bone's internal make up (its mineralization and density) and its shape are dictated by the bone's function, and normal bone function is characterized by continually adapting to the environment.

Although the primary influence on the architecture and the materials that make up the bone is genetic, adaptations are constantly being made. These adaptations are based on two things: the person's nutrition and the amount and kind of stresses that the bone is subjected to.

Types of forces

Normal weight-bearing and everyday activities, like walking and standing, exert a variety of forces on the bones, including compression, tension, torsion, and shearing:

- ✓ **Compression:** *Compression* refers to the squeezing force that occurs through normal weight bearing, as your body weight and gravity push down on your frame.

- ✓ **Tension:** *Tension,* which happens during normal weight-bearing and other physical activities, occurs when a muscle or tendon pulls on a bone at its attachment site to create movement or increase stability. Extending your knee, for example, creates tension on the tibial tuberosity (refer to the preceding section).

- ✓ **Torsion:** *Torsion* is a twisting effect. When you're standing and twist to change directions, torsion is exerted on the tibia (leg bone). This twisting is usually combated by the strength in the bone, but sometimes the bone may break.

- ✓ **Shearing:** *Shearing* refers to the tearing across the longitudinal axis. When you stop abruptly and your foot extends in front of you to stop your forward momentum, your leg bone experiences a shear force. Part of the bone wants to continue going in the same direction as your body's momentum, but the foot stopping your forward movement pushes it back.

The modeling threshold

Each bone has a *modeling threshold,* which determines how much and what kind of force can be applied to the bone before the bone begins to adapt. A bone is able to accommodate a certain type and amount of force; when forces equal the modeling threshold in both direction and amount, no change in bone structure occurs. In essence, the bone has been uniquely designed to handle that activity.

Changing the type or intensity of an activity results in forces exceeding the threshold and stimulating bone development. The development of new bone can enhance *mineralization,* the addition of *osteocytes* (bone cells). When forces delivered to the bones decrease and produce less demand, bone mineralization lessens, and osteocytes are reabsorbed. (**Note:** *Osteoblasts* and *osteoclasts* are two types of osteocytes.)

Increases in mineralization occur only in the areas that are experiencing the applied forces. For example, runners often have higher bone density in their lower extremities than in their upper extremities. Likewise, baseball and tennis athletes often demonstrate increases in their throwing or racquet arms while their lower extremities are no different from other athletes' lower extremities. The same is true for most active folks in their dominant versus non-dominant sides. The areas that experience the greater forces typically have greater bone diameters, cortical widths, densities, and calcium concentrations.

Knowing the kind of old bone you are

Typically, the skeletal system is divided into two parts, the appendicular skeleton and the axial skeleton, shown in Figure 8-2:

- ✔ **Axial skeleton:** This skeleton is made up of the skull, vertebrae, sternum, and ribs. The axial skeleton is considered the central aspect of the skeletal system.

- ✔ **Appendicular skeleton:** This skeleton is made up of the bones that form the appendages.

In addition to being considered part of either of these two skeletons, bones are further characterized by their shapes. When you look at the skeleton, you see that bones are shaped very differently — short, long, flat, irregular, and sesamoid. As we note earlier, the specific function that the bone plays within the body dictates its shape and size. This list has the details:

- ✔ **Short bones:** These bones are small, solid, and often cube shaped. They have a relatively large articulating surface and are able to articulate with more than one bone and typically contribute to gliding. Short bones include the carpals (wrist bones) and tarsals (ankle bones).

- ✔ **Flat bones:** As the name suggests, these bones are flat or slightly rounded and vary in thickness. Their job is to protect organs within the body. They include the ribs, sternum, ilium (the bone forming the upper part of the pelvis), clavicle (collar bone), and scapula (shoulder blade).

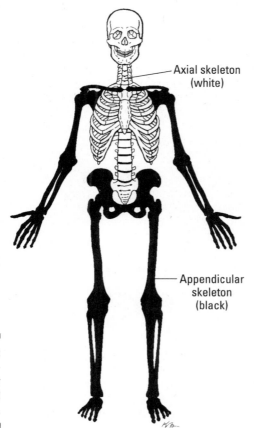

Axial skeleton
(white)

Appendicular
skeleton
(black)

Figure 8-2:
The appen-
dicular
and axial
skeleton.

✔ **Irregular bones:** Irregular bones play a number of different roles, depending on the nature of their irregularities. Typically the irregularities are
dictated by muscle and other soft-tissue attachments that result in forces
being applied. For example, the vertebrae (the small bones of the spine)
possess a tunnel through which the spinal cord travels and also have a
number of exterior protrusions where muscle, tendons, and ligaments
attach. Irregular bones include the vertebrae, coccyx (tailbone), ischium
and pubis (other bones in the pelvis), and sacrum (part of your tailbone).

✔ **Long bones:** Long bones are the main support of the appendicular
skeleton. They're typically elongated and cylindrical. Given their length
and the processes and protuberances where muscles, tendons, and
ligaments attach, they play a role in the lever system (read more about this
in Chapter 7). Additionally, the long shafts house a medullary canal, which
holds bone marrow, and articular cartilage coats the bone's end to protect
it from wear and tear. Examples of long bones include the radius and ulna
(bones in the forearm), and the fibula, femur, and tibia (leg bones).

✔ **Sesamoid bones:** These bones are embedded within various tendons. Their role is to protect and to enhance mechanical advantage of the muscle-tendon unit. The patella (kneecap) is the largest and most recognized sesamoid bone.

Growing Up is Hard to Do: Examining Bone Growth

Your bones begin as cartilaginous masses when you're a fetus, and progressively morph into the form you're familiar with today. Bone development continues through your life, as more bone is either added or taken away. Both the loss and gain of bone cells are normal occurrences throughout your life.

The long and the short of it: Longitudinal and circumferential growth

Bone growth occurs in two distinct ways. *Longitudinal growth* happens as a bone lengthens, and *circumferential growth* refers to its increase in diameter. Here are the details (see Figure 8-3):

✔ **Longitudinal growth:** This type of growth occurs in a longitudinal direction and happens at the *epiphyses* (the large, knobby ends of long bones). At the location of the epiphyseal plate, the diaphysis continually produces new bone cells, resulting in growth. Longitudinal growth continues through adolescence, until the epiphyseal plates close and are no longer able to contribute. Typically, growth ceases during the later teens, although some growth may continue a few years beyond that.

✔ **Circumferential growth:** Here, the bone grows in diameter. In circumferential growth, concentric layers of bone develop around what's already there. Typically, the concentric layers of growth begin around the inner layer of the periosteum and continue to build upon that. As the layers of bone are being laid down on the outer layer (the *periphery*), the inside of the bone is being reabsorbed. This reabsorption allows for the medullary cavity to continue to grow proportionally and stops the bone from becoming too rigid and heavy (which would happen if the inside bone cells didn't get reabsorbed).

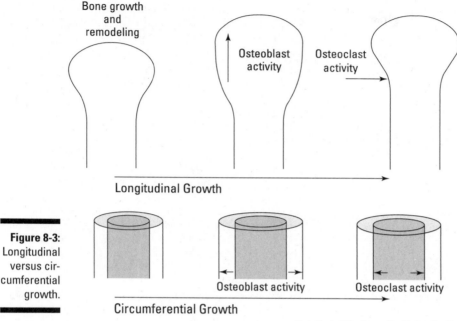

Illustration by Wiley, Composition Services Graphics

Figure 8-3:
Longitudinal
versus cir-
cumferential
growth.

Bone growth
and
remodeling

Osteoblast
activity

Osteoclast
activity

Longitudinal Growth

Osteoblast activity

Osteoclast activity

Circumferential Growth

Both types of growth are dependent on specialized cells called *osteocytes*. Two types of osteocytes are responsible for the modeling and remodeling that occurs through the lifespan. *Osteoblasts* are responsible for forming new bone, and *osteoclasts* resorb old bone. The tireless activity of these cells in response to use and disuse are what makes up bone modeling and remodeling.

Increases and decreases in density

As we note in the earlier section "Pouring over porosity: Cortical versus trabecular bones," a bone's density is related to its mineralization and the stresses that are applied to it. Hormones also influence bone density.

During adolescence

During childhood and adolescence, bone *hypertrophy* occurs in response to activity. That is, bone density increases fairly rapidly as the bones continually *ossify* (harden and become more dense) to meet demands of the structural system as children walk, run, and otherwise function in their daily activities.

During adulthood

Mineralization dictates a bone's strength and ability to accept forces. In the adult years, typical bone development and modeling occurs as bone mineralization takes place in response to forces. As forces increase or are maintained, the more mineralized bone becomes. Conversely, as fewer forces are delivered to the bone, the bone becomes less mineralized and less able to accept stresses.

Although being active can result in increased bone density, through the adult years, beginning in the third or fourth decade of life, a decrease in bone density or demineralization occurs naturally. For women, typically beginning in their mid-20s, and for men, nearly a decade later, bone density typically begins to fall at a progressive rate. As the bone's density decreases, the materials that make up the bone decrease as well. As the dismantling of the network that was the trabeculae occurs and the bone becomes more porous (refer to the earlier section "Pouring over porosity: Cortical versus trabecular bones"), the strength of the framework is significantly compromised. Bones become progressively weaker and less resilient to the forces they once successfully resisted.

Unfortunately, the loss in bone mineralization has been shown to affect women to a larger extent than men. In fact, women lose approximately 0.5 percent to 1 percent of their bone mass each year until menopause. After menopause, bone mass loss increases to a much larger degree. Decreases often are most apparent in these groups:

- **Those who are not physically active:** People who are sedentary, bedridden, or not able to ambulate in a way that puts stress on the bones suffer from demineralization.

- **Those who are not affected by gravity:** Yeah, we're referring to astronauts, especially those living at the international space station for extended periods of time.

During old age

Demineralization is problematic for anyone because when bone density decreases the bone's ability to resist and accommodate imposed stresses and forces also decreases. People suffering from demineralization tend to experience more fractures, many of which are quite debilitating.

Demineralization is a significant issue in the elderly. First, beginning in a person's 20s and 30s, a progressive loss of mineralization is natural. Additionally, bones' tensile strength and elasticity are also negatively impacted as a person grows older. Some studies have shown that physical activity through the lifespan can help to thwart bone demineralization in men and women, yet the inactive elderly often experience osteoporosis.

Limb lengthening: Getting evened out

Sometimes the length of the limbs may not be equal, or the limbs may be extraordinarily short. A number of syndromes or conditions result in shortened or shorter-than-average limbs. Sometimes the limbs don't grow equally; other times, misalignments may exist as a result of injury.

When such conditions exist, physicians may recommend *distraction osteogenesis* (limb lengthening) as a solution. This procedure can be applied to long bones and involves the physician fracturing the bone and slowly separating, or *distracting,* the fracture site over time. Distracting the fracture site slowly over time stimulates healing, and new bone grows to fill the void. In addition to stimulating bone growth in that area, the procedure also stimulates the soft tissues, like the blood vessels and muscles, to accommodate the growth and catch up. The patient or doctor increases the separation, typically 1 mm a day over a series of weeks.

Distraction osteogenesis is a tricky procedure: If the separation is too great, growth doesn't occur, and a permanent separation results. In addition, if the space isn't separated soon enough between stages of the lengthening process, the bone solidifies, preventing the two

ends from being moved without being broken again. Ouch!

Although the most common bone involved in limb lengthening is the femur, the procedure has been performed on a number of different long bones, like the humerus, tibia, and fibula. Also, some maxillofacial surgeons (surgeons who work on the jaw and face) use this type of a technique to encourage changes within the jaw.

The lengthening procedure isn't without its drawbacks. First, it's a surgical procedure, and there are always inherent risks with any surgery. Second, this procedure involves the controlled breaking of the bone and a device being applied that assists with the future separation. It also takes several weeks to complete, depending on how much lengthening is needed or wanted. Interestingly, one issue doesn't seem to be excessive pain. Despite sounding a lot like a torture device, reports indicate that patients don't seem to experience significant pain or discomfort throughout the procedure.

Obviously, this procedure isn't for everyone, but it has been shown to increase the length of a long bone anywhere from 15 percent to 100 percent of its original length, and despite the controllable risks for infection, it has great outcomes.

Examining Osteoporosis

Demineralization is a fact of life for many but in particular for senior citizens and female athletes. It can lead to *osteoporosis,* a disorder involving decreased bone mass and strength that result in one or more often painful and debilitating fractures. Figure 8-4 shows the difference between a healthy bone and a bone with osteoporosis.

Normal healthy bone

Bone with osteoporosis

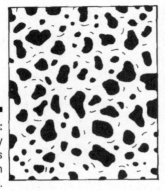

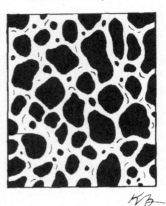

Figure 8-4:
A healthy
bone versus
a bone with
osteoporosis.

Illustration by Kathryn Born, MA

Osteoporosis begins as a condition called *osteopenia,* which is characterized by reduced bone mineral density that predisposes individuals to bone fractures. Basically, people with osteopenia have a marked level of bone demineralization but haven't experienced a fracture yet. Often osteopenia progresses to osteoporosis.

Osteoporosis is considered a serious health risk for both men and women. Nearly half of all elderly postmenopausal women and one-third of all men (elderly included) will suffer from osteoporosis. As the average age of these groups rises, so does the risk of developing this condition. Identified risk factors include being female, being of white or Asian ethnicity, being older, having a small body type, and having a family history of the condition.

Types of osteoporosis

Osteoporosis is characterized into two separate types: postmenopausal (also called type I), which affects only women, and age-associated (type II), which affects both men and women. Another population directly impacted by osteoporotic conditions later in life is that of the female athlete, many of whom are susceptible to osteoporosis because of the *female athlete triad.*

Type I osteoporosis

Type I osteoporosis, otherwise known as *postmenopausal osteoporosis,* affects a large proportion of women over 50. The steady decline in bone density of older women coupled with a significant increase in the progression of this loss after menopause creates significant implications for this group. Osteoporotic fractures typically begin around 15 years after menopause and represent about three times as many femoral, neck, and vertebral fractures as men of the same age.

Bone density in men and women

Over the years, a significant amount of research has occurred regarding bone mineralization and density. In particular, researchers have explored the differences between men and women. Men typically possess greater bone densities in comparison to women. Following are the possible reasons why:

✔ **Bone density is determined largely by weight, body fat, exercise intensity, and hormonal influences.** Each of these factors contributes in the grand scheme to why men have greater densities than women.

✔ **The hormone estrogen has been linked to preserving bone density in women; unfortunately, estrogen depletion is all too common in women as they age.** In the earlier section "Examining Osteoporosis," we discuss osteoporosis, which may result from loss of estrogen, among other things. Other groups at increased risk for estrogen reduction are women who are *amenorrheic* (don't have periods). ***Note:*** Although amenorrhea can be caused by a number of factors, a common one is excessive physical activity.

Type II osteoporosis

Also called *age-associated osteoporosis,* type II exists in both men and women over 70. A significant number (90 percent) of the fractures occur after age 60 and are often very serious, representing one of the leading causes of death in the elderly.

You may have noticed that, as your grandparents age, they seem to shrink. Well, that actually *is* what's happening, and it's happening in a painful way: Essentially, they're experiencing multiple fractures of the vertebrae over time. Here's how it works: As osteoporosis progresses, the trabecular network decreases, resulting in crush fractures caused by the weight-bearing activities of daily living. These crush fractures typically occur toward the anterior aspects of the vertebrae and result in a forward — or "hunchback" — posture.

The increase in prevalence of osteoporosis in both men and women is occurring at a staggering rate. Some reports have indicated a three- to four-times increase in the last three decades. Despite the progressive and unavoidable decrease in bone density that occurs across the lifespan, don't discount the importance of being active early and often throughout your life. Doing so may help lessen these affects later on.

Female athlete triad

Athletes are very competitive and are driven to excel in their events. Female athletes in particular experience what is called the *female athlete triad,* which because of its characteristics, is often difficult to recognize and treat. In this triad, an athlete experiences disordered eating habits, amenorrhea, and osteoporosis, which later in their lives becomes advanced osteoporosis.

Together these three interrelated conditions are significant contributors to osteoporosis and unhealthy scenarios for young women. Here's what needs to be done: First, during an athlete's younger years, when a significant amount of bone density needs to be created, coaches and parents must ensure that that young athletes participate in a way that maximizes their bone mineralization. Second, care should be taken to educate female athletes on the importance of positive body image and self-confidence, normal menses, and appropriate nutritional habits.

Of course, although a majority of these issues are directly related to the female athlete, similar issues with body image and disordered eating has been documented in males as well. In both cases, these athletes should be monitored and cared for appropriately.

Bone up! Preventing osteoporosis

Although a nutritious diet, regular exercise, and normal hormone levels have all been shown to facilitate bone density increases, the most effective treatment is prevention. The single most important factor in avoiding osteoporosis or limiting its affects later in life is to build a high peak of bone mass early in childhood and adolescence. Here's how:

- ✔ **Stay (or get) active.** Resistance training, running, walking, jumping, and a whole host of low- and high-impact activities have been shown to have a positive influence on bone mineralization. Engage in an exercise routine that's appropriate for your overall health; a simple consultation with your physician may help in this regard.

- ✔ **Check your hormone levels.** Hormone levels are associated with influencing bone mineralization. For instance, estrogen in women and testosterone in men have been shown to assist in maintaining bone density. You can discuss your hormone levels with your physician.

- ✔ **Get enough calcium.** Those who have a dietary deficiency of calcium may benefit from calcium supplementation (supplemented with Vitamin D, which assists with calcium absorption).

- ✔ **Guard against other risk factors for osteoporosis.** These include tobacco smoking, weight loss, and consuming too much caffeine or protein (both of which are thought to contribute in some way to progressing osteoporosis). Consult with your healthcare provider, who can better assess your individual risk for osteoporotic conditions in the future.

Making a Break! Bone Fractures

The skeletal system has significant impact on your ability to perform activities of daily living, which may include something as simple as walking to the refrigerator or lifting up your purse. Your bones bear not only the weight of your body but also the additive forces that accommodate many of these everyday functions. No wonder injuries occur! Chances are you know of someone who's broken a bone, or you've experienced that injury yourself.

Investigating types of fractures

Fractures, euphemistically (or officially) defined as "a disruption in the continuity of the bone," are the most common type of bone injury. Often referred to as "breaks," fractures can be debilitating and certainly very painful. Some fractures are minor, requiring only minimal treatment; others are very complex and require surgical treatment.

The kind and severity of the injury you end up with is dependent on several factors: the type, direction, amount, and timing of the force(s) delivered to the bone, as well as the maturity, nutrition, and hydration of the bone.

In a *simple fracture,* the bone remains within the skin; with a *compound fracture,* the bone exits the skin. Both are painful because the periosteum, the outer covering of the bone, is highly innervated with nociceptors (refer to the earlier section "The structural components of bone" for info about nociceptors). Read on to find out about the various types of fractures some of which you can see in Figure 8-5.

Figure 8-5:
Types
of bone
fractures.

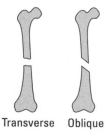    

Transverse   Oblique   Spiral   Comminuted   Avulsed   Impacted   Greenstick

Illustration by Wiley, Composition Services Graphics

- **Transverse:** A *transverse fracture* is one that occurs perpendicular to the long axis of the bone. Often transverse fractures are caused from a direct blow. Common mechanisms of injury for this type of fracture are motor vehicle accidents or high impact sports like football or rugby. This type of fracture usually becomes displaced and needs to be realigned to ensure proper healing. Commonly, these types of fractures are realigned with what is called an *open procedure,* in which the bone is surgically *reduced;* that is, the doctor uses plates and screws to keep the bone together before splinting or casting the limb.

- **Oblique:** An *oblique fracture* is one that occurs at a diagonal to the shaft of the bone. The diagonal orientation usually crosses only one plane (that is, it's a straight fracture at an angle). An oblique fracture is usually caused by a combination of bending and twisting across the shaft of the bone. This type of fracture also needs to be treated immediately, and surgery may be needed to realign this fracture.

- **Spiral:** A *spiral fracture* is often confused with an oblique fracture. However, a spiral fracture is very different in that it spirals around the shaft of the bone. Unlike the oblique fracture, which involves only one plane, the spiral fracture spans multiple planes. This type of fracture occurs when the bone is forced to twist with enough force to cause the bone to fail. An example would be someone planting one foot really hard while changing direction when running; the force is so great when the person tries to change direction that the bone torques and fractures.

- **Comminuted:** A *comminuted fracture* is one that causes the bone to break into more than two fragments; typically the bone is splintered or crushed. This type of fracture can be caused by a large compressive force but more often is the result of multiple types of forces coming from different directions. These fractures nearly always require surgery to repair.

- **Impacted:** An *impacted fracture* occurs when one broken end of the bone is wedged into the other broken end. Different from a comminuted fracture because the bones become wedged into each other, this type of fracture is often the result of significant axial compression, typically caused from falling a distance to the ground. Like a comminuted fracture, an impacted fracture often, if not always, requires surgery.

- **Greenstick:** Children have softer, more flexible bones than adults, and their bones are more apt to bend than break completely. A *greenstick fracture* occurs when a bone cracks but doesn't break all the way through. These fractures can be difficult to diagnosis because they often don't present like a typical fracture; that is, they aren't displaced, and the amount of pain varies greatly. Regardless, all fractures — including incomplete ones like greenstick fractures — must be immobilized to ensure proper healing.

- **Avulsion:** Fractures that involve a piece of bone being pulled away by a previously attached ligament or tendon are called *avulsion fractures.* Avulsion fractures can occur anywhere in the body but are most common at the pelvis because of the amount of force that is transferred through the large muscles that attach to it. This type of injury is more common in children than adults because children's bones are softer and more flexible and fail before the ligament or muscle tears. (In adults, the ligament or muscle tendon is more likely to fail before the bone because of the higher degree of bone mineralization and large cortical bone makeup.)

✔ **Stress:** Low magnitude forces can sometimes lead to stress fractures. Typically a *stress fracture* involves a small disruption of the cortical layers of the bone, but the bone doesn't actually break (although a complete cortical fracture can develop if force is continually applied to this part of the bone). Typically, these types of fractures are seen in people who engage in repetitive activities like running. When the bone experiences repetitive loading at a large enough frequency and magnitude, it responds with a stress reaction. The stress reaction facilitates bone remodeling because of the damaged tissue created from the overload of stress at the site. As with all other fractures, eliminating the source of the fracture is essential and immobilization is needed. Typically, these types of fracture don't require surgery to repair.

With stress fractures, bony deformation doesn't occur, but the bone is damaged because of the repetitive micro-trauma. A common reaction to the repetitive micro-trauma involves osteoclasts being delivered to the area to absorb the damaged tissue, while osteoblasts follow to deliver new bone to the site. When the bone isn't able to go through a full stress reaction that involves bone absorption and delivery, the cumulative effect is a stress fracture.

✔ **Epiphyseal:** Commonly referred to as *growth plate fractures,* epiphyseal fractures occur to the cartilaginous epiphyseal plate (refer to the earlier section "The structural components of bone" for more info on the epiphyseal plate). Both acute and repetitive loading can result in a growth plate fracture. This type of fracture may result in premature closure of the epiphyseal junction and can significantly affect bone growth.

Remodeling after a fracture

After a fracture occurs, the osteocytes — specifically, the osteoblasts and osteoclasts — kick into gear to help heal the bone. Here's what happens (see Figure 8-6):

1. **A bony callus forms.**

   Osteoblasts are recruited to the location to form a callous. This callous serves to splint and support the fracture site. Over time, typically four to six weeks (about how long you need to wear a cast), the osteoblasts work to build up the fracture site again.

2. **After a strong bony callus has formed and initial ossification has occurred, the bone begins to remodel.**

   During remodeling, osteoclasts are sent to the site of the new bone to begin to break the bone back down to how it was before it was ever broken. The remodeling can take several months before the bone resumes its original shape and strength.

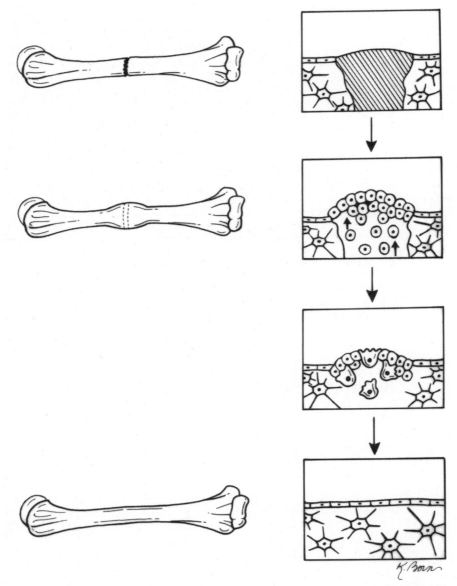

Figure 8-6:
The role of osteo-blasts and osteoclasts in bone remodeling.

Illustration by Kathryn Born, MA

When we all fall down!

Getting older just isn't fair. The Centers for Disease Control and Prevention (CDC) reports that every year one in three adults age 65 and older experiences a falling episode. These falls are moderately to severely debilitating and can increase the risk of early death. Falls account for the most common cause of nonfatal injuries and hospital admissions.

Both men and women are at risk of falling, with men experiencing 34 percent more falls than women. Interestingly, falls rates seem to differ by ethnicity, with older whites being 2.4 times more likely to die from falls than blacks, and older non-Hispanics experience a higher fatal fall rate than Hispanics. When examining for gender, it's very clear that, although men are more likely to die from a fall, women are twice as likely to have a fracture as a result of a fall than men are — a tendency that relates to the earlier discussion about osteoporosis and women

(see the section "Examining Osteoporosis"). Another interesting fact presented by the CDC is that 95 percent of hip fractures are caused by falls, accounting for 271,000 episodes in 2009.

Given the overwhelming number of falls that occur in a year in the U.S. alone, it's only natural to wonder what can be done to minimize this epidemic. The primary recommendation: Move always and always move. Many of the imbalances and slowing of reflexes that cause falls can be avoided simply by remaining physically active throughout life. Of course, just being active isn't the only thing you can do. Other things include making your home more safe by reducing the tripping hazards (grandkids' toys, having pets underfoot, and so on), having your medications checked (some contribute to dizziness or drowsiness), and having your glasses checked (if you can't see where you're going . . .).

Chapter 9

These Joints Are A-jumping!

. . . .

In This Chapter

▶ Identifying types of joints and their functions

▶ Understanding joint stabilization

▶ Investigating flexibility and stretching techniques

. . . .

*W*ithout the joints that make up your body, you wouldn't be able to move. Certainly, movement is dependent on your muscles' ability to create the forces that propel you and your bones' ability to provide the necessary structural support. But imagine trying to walk up stairs while keeping your knees straight or trying to write without bending your fingers or wrist. Life as you know depends on your ability to bend and twist and glide.

Movement is largely dictated by your anatomical makeup, but things like flexibility and prior history of injury, which differ from person to person, can have significant impacts on your activities and ability to avoid injury. The purpose of this chapter is to investigate the types of joints that are in the body, explain how stability is established, and delve into the influence that flexibility has on the tasks you perform daily.

Getting These Old Bones to Move: Types of Joints

Just as the body has different types of bones (refer to Chapter 8), the body also has different types of joints. You're aware of these differences just by observing how your body moves or what its limitations are: Have you ever wished that you could spin your knee similarly to your neck or tried to extend your back only to realize you've gone too far? Have you ever wondered why some joints move one way and others go another way? These differences are all based on anatomical design and joint architecture. Joint architecture dictates how bones move in relation to one another and the range of motion that's created as a result.

Not all joints allow for the same range of motion. In fact, some joints don't even move at all, whereas others move freely. Joints can be characterized by how much they move, in what directions, and to what extent. Additionally, they can be classified by structure and function, and by the number of axes of rotation that they allow. The following sections have the details.

A joint is a joint, but in the study of joints, you need to know and be comfortable with their technical name: *arthrosis* (singular) and *arthroses* (plural). Why? Because the terms used to classify joints by the way in which they move are based on this technical term, not the common one: *synarthroses, diarthroses,* and so on. Remembering these terms and what they mean will just be easier if you latch onto *arthroses* now. To help you out, we use the technical and common terms interchangeably throughout this section.

Structural classifications: Fibrous, cartilaginous, and synovial

Structural classifications of the arthroses include three types: *fibrous, cartilaginous,* or *synovial:*

- ✔ **Fibrous joints:** In these joints, fibrous tissue or cartilage connects the bones. Fibrous joints are slightly movable. An example is the lower arm (the radio-ulnar joint).

- ✔ **Cartilaginous joints:** These joints contain cartilage, either of the hyaline or fibrocartilage variety. They move a bit more than fibrous joints, and examples include the pubis (a type of bone in the pelvis) and vertebrae (the small bones in your spine).

  Hyaline cartilage, the most common kind of cartilage, generally covers the articular surfaces of synovial joints (see the next item in this list), where it reduces friction, protects against shock, and allows the joint range of motion. *Fibrocartilage* is nice and spongy, which makes it a good shock absorber. You'll find it between the vertebrae of the spine, for example. For detailed information on cartilage, head to the later section "Enhancing Joint Stability and Longevity: Cartilage and Connective Tissues."

- ✔ **Synovial joints:** These are the most common kind of joint. These joints have a cartilaginous covering (hyaline cartilage) where the articulating bones meet and a *synovial cavity* (also called a *joint capsule*), which is essentially a space between the bones of the joint, where a collection of soft tissues provide stability and synovial fluid keeps everything nice and lubricated. Synovial joints possess the greatest range of motion. Examples include the elbow, wrist, hip, and shoulder joints.

Functional classifications: Synarthroses, diarthroses, and more

Classifying the joints by function allows you to separate them by how they work.

Immovable joints: Synarthroses

Synarthroses are joints that don't move. Examples of synarthroses include the following:

- ✔ **Sutures:** This kind of joint doesn't allow for any movement. The joints between the bones of the skull are examples of sutures. The irregularly shaped bones of the skull are joined very tightly and don't move.

- ✔ **Syndesmosis joints:** A *syndesmosis* is a joint that is connected by a significant amount of dense, fibrous tissue. The dense tissue surrounding this joint allows for only very limited motion. You find syndesmosis joints very tightly connecting the radius and ulna (called the *radio-unlar joint* in the lower arm) and the tibia and fibula (called the *tibio-fibular joint,* in the ankle).

Slightly movable joints: Amphiarthroses

Amphiarthroses are slightly movable joints connected by fibrocartilage or hyaline cartilage. Examples include the intervertebral disks, which lie between each vertebra and allow slight movement from one segment to the other. These types of joints help to absorb forces (like compression in the spine) and allow for more motion than the synarthroses.

A type of amphiarthrotic joint is the symphysis. The *symphysis* is a joint that is separated by a fibrocartilagenous disk or a very strong ligament that links two bones together. Examples of amphiarthrotic joints are the pubic symphysis of the pelvis, the spine, and the joint that connects the scapula to the clavicle in the shoulder. Each allows for only a little movement while providing support and stability for the bones they're attached to.

Freely movable joints: Diarthroses

Diarthroses, which are joined together by ligaments, are the most common type of joint in the body. They're also called *synovial joints* (refer to the earlier section "Structural classifications: Fibrous, cartilaginous, and synovial") because a cavity between the two connecting bones is lined with a synovial membrane and filled with synovial fluid, which helps to lubricate and cushion the joint. The ends of the bones are cushioned by hyaline cartilage.

Table 9-1 lists the several different types of diarthrotic joints and describes the kinds of movements each type allows. You can also see these joints in Figure 9-1.

Table 9-1	Types of Diarthrotic Joints		
Type of joint	**Description**	**Movement**	**Example**
Ball-and-socket	The ball-shaped head of one bone fits into a depression (socket) in another bone.	Circular. The joints can move in all planes, and rotation is possible.	Shoulder, hip
Ellipsoid joint	An oval-shaped protuberance (called a *condyle*) of one bone fits into oval-shaped cavity of another bone.	Can move in different planes but can't rotate.	Knuckles (the joints between metacarpals and phalanges)
Gliding joint	Flat or slightly curved surfaces join together.	Sliding or twisting in different planes.	Joints between carpal bones (wrist) and between tarsal bones (ankle)
Hinge joint	The convex surface of one bone joins with concave surface of another.	Up and down motion, bending *(flexion)*, and straightening *(extension)*.	Elbow, knee
Pivot joint	Cylinder-shaped projection on one bone is surrounded by a ring of another bone and ligament.	Rotation is the only movement possible.	Radio-ulnar joint at the elbow (supination/pronation), atlanto-axial joint of the neck under the head (head rotation)
Saddle joint	Each bone is saddle shaped and fits into the saddle-shaped region of the opposite bone.	Many movements are possible; can move in different planes but can't rotate.	Carpo-metacarpal joint of the thumb

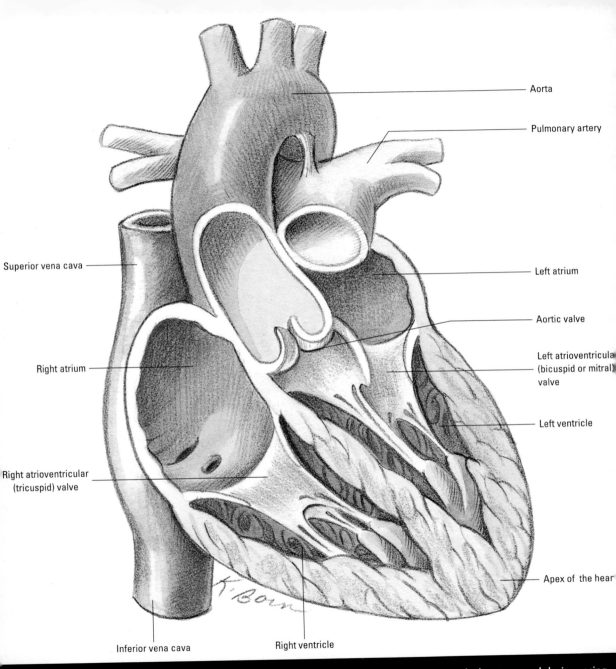

Aorta

Pulmonary artery

Superior vena cava

Left atrium

Aortic valve

Left atrioventricular (bicuspid or mitral) valve

Right atrium

Left ventricle

Right atrioventricular (tricuspid) valve

Apex of the heart

Inferior vena cava

Right ventricle

The heart is responsible for directing blood to the lungs and the rest of the body. During exercise, the heart responds by increasing

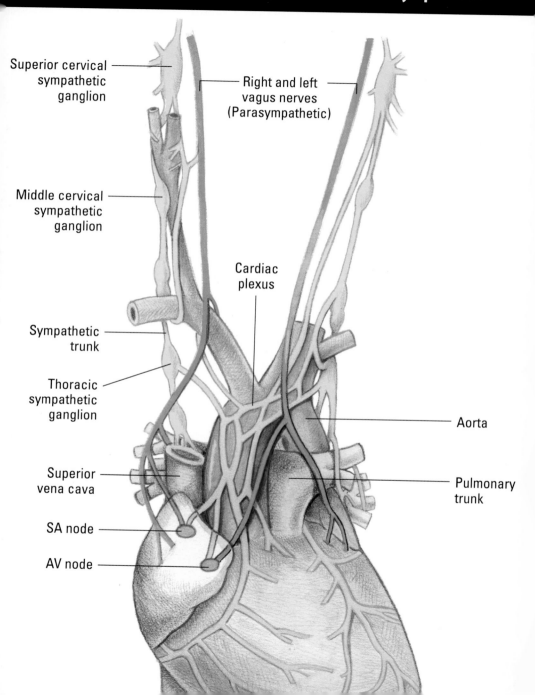

Superior cervical sympathetic ganglion

Right and left vagus nerves (Parasympathetic)

Middle cervical sympathetic ganglion

Cardiac plexus

Sympathetic trunk

Thoracic sympathetic ganglion

Aorta

Superior vena cava

Pulmonary trunk

SA node

AV node

Core-to-Shell Heat Transfer

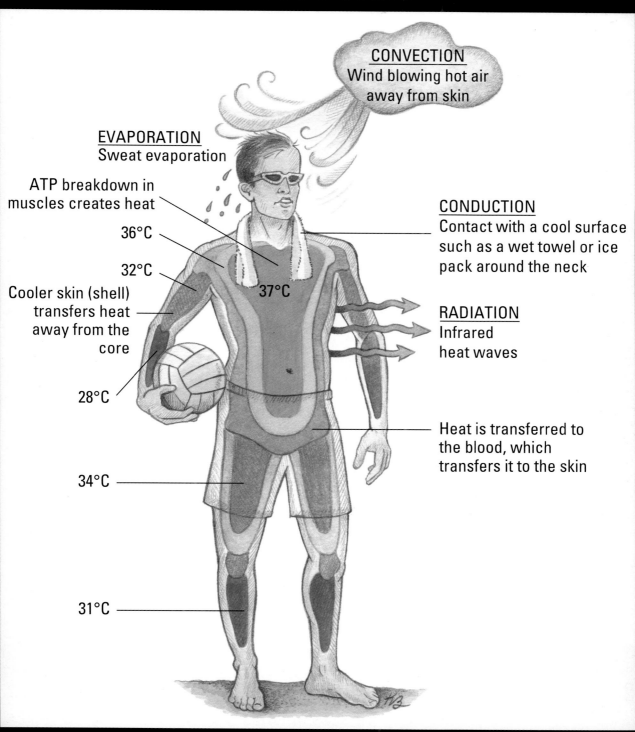

CONVECTION
Wind blowing hot air away from skin

EVAPORATION
Sweat evaporation

ATP breakdown in muscles creates heat

36°C

32°C

Cooler skin (shell) transfers heat away from the core

28°C

34°C

31°C

37°C

CONDUCTION
Contact with a cool surface such as a wet towel or ice pack around the neck

RADIATION
Infrared heat waves

Heat is transferred to the blood, which transfers it to the skin

The breakdown of ATP generates large amounts of heat within the body core. The heat is transferred from the muscles outward to the skin for dissipation.

The Many Ways of Cooling the Skin

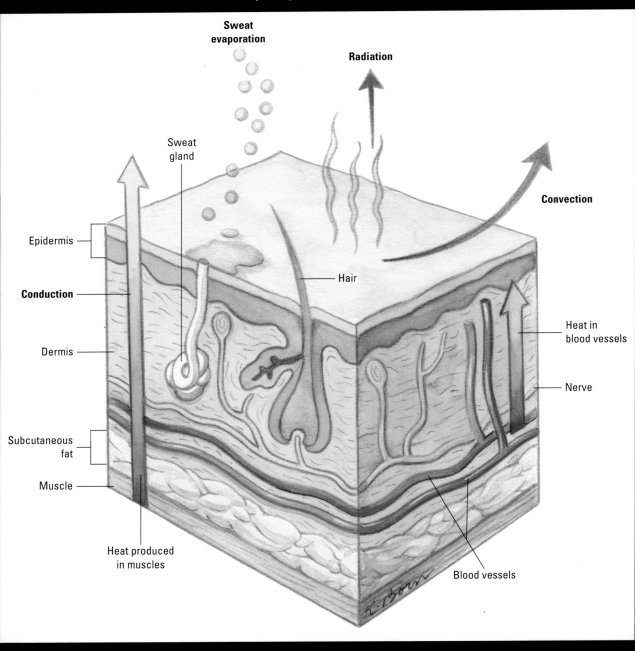

Sweat evaporation

Radiation

Convection

Sweat gland

Hair

Epidermis

Conduction

Heat in blood vessels

Dermis

Nerve

Subcutaneous fat

Muscle

Heat produced in muscles

Blood vessels

Heat is transferred from the core of the body outward to the skin. Upon arrival at the skin, the heat is taken away from the body in a variety of ways.

Anatomical Position

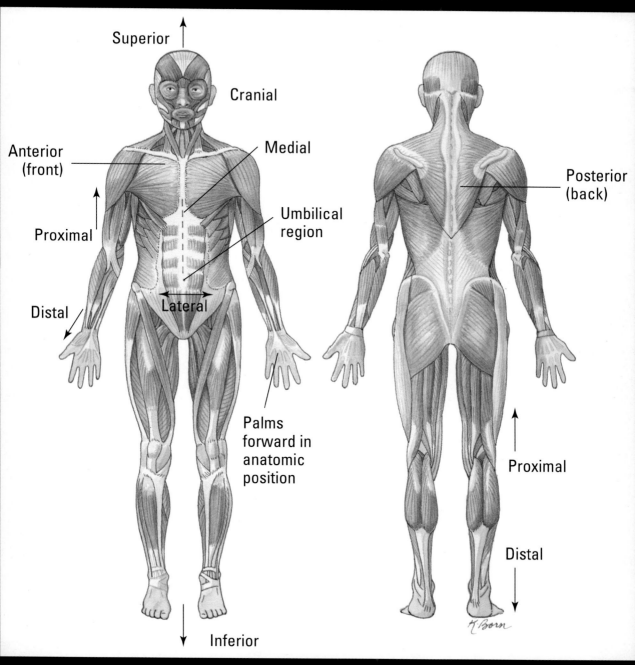

Superior

Cranial

Anterior (front)

Medial

Proximal

Umbilical region

Distal

Lateral

Palms forward in anatomic position

Posterior (back)

Proximal

Distal

Inferior

K. Born

Looking beneath the skin, you can see that the underlying muscles are intertwined and complex

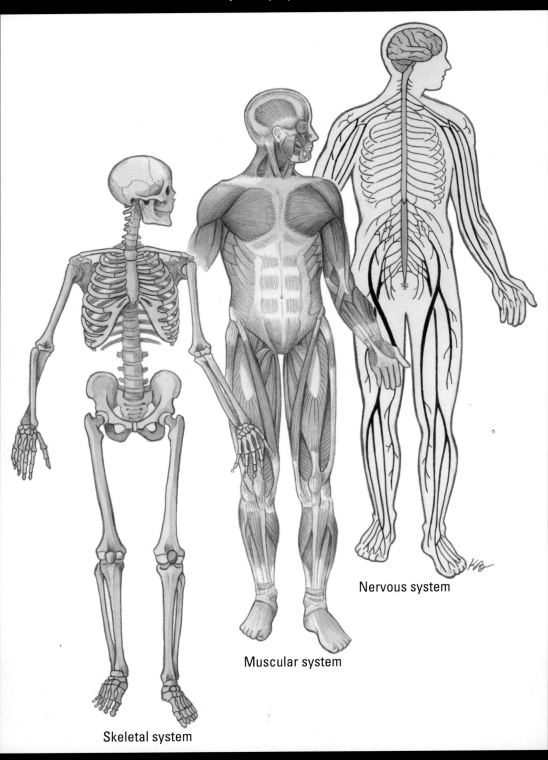

Nervous system

Muscular system

Skeletal system

The different systems of the body perform their unique functions while collectively working together. The nervous, skeletal, and muscular systems work together and independently to help you move.

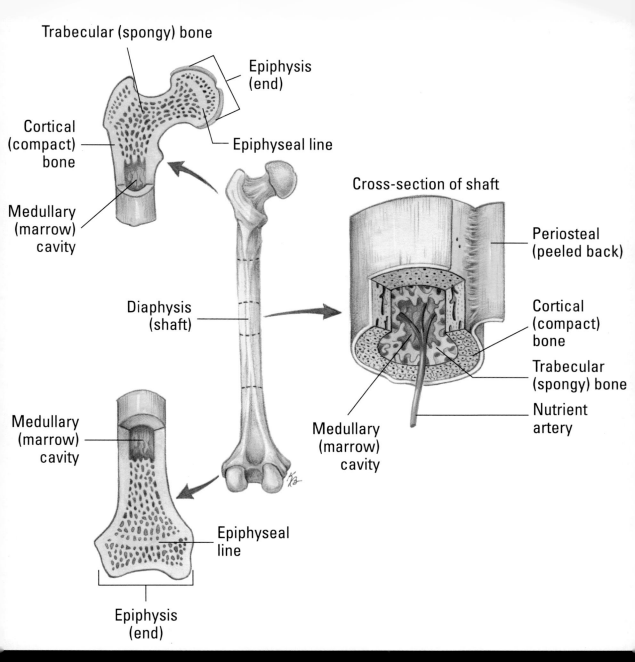

Trabecular (spongy) bone

Epiphysis (end)

Epiphyseal line

Cortical (compact) bone

Medullary (marrow) cavity

Cross-section of shaft

Periosteal (peeled back)

Cortical (compact) bone

Trabecular (spongy) bone

Nutrient artery

Medullary (marrow) cavity

Diaphysis (shaft)

Medullary (marrow) cavity

Epiphyseal line

Epiphysis (end)

The skeletal system is responsible for providing the support needed to make movement possible

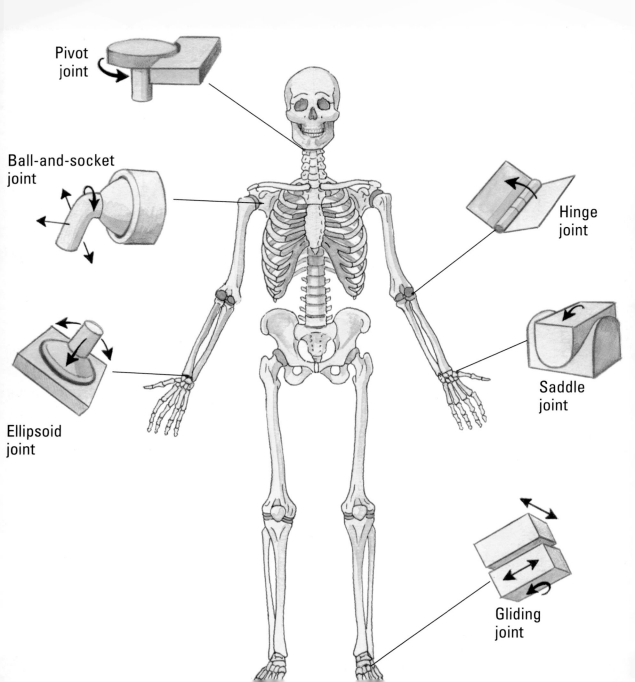

Pivot joint

Ball-and-socket joint

Ellipsoid joint

Hinge joint

Saddle joint

Gliding joint

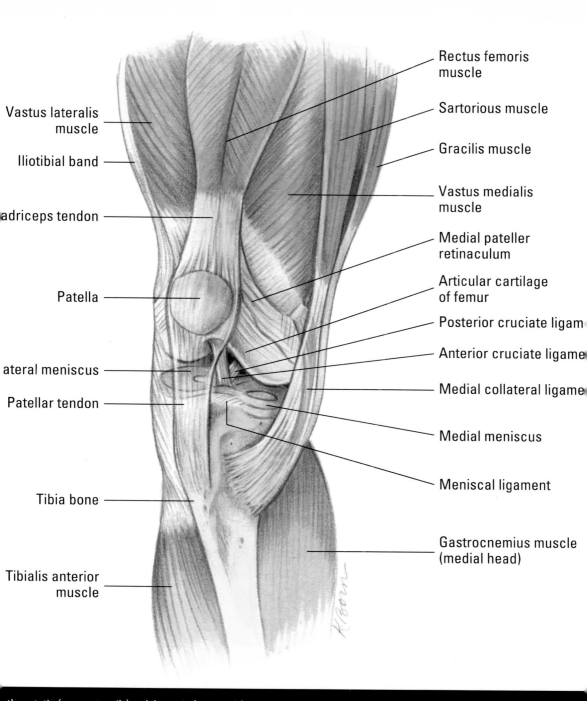

Vastus lateralis muscle

Iliotibial band

adriceps tendon

Patella

ateral meniscus

Patellar tendon

Tibia bone

Tibialis anterior muscle

Rectus femoris muscle

Sartorious muscle

Gracilis muscle

Vastus medialis muscle

Medial pateller retinaculum

Articular cartilage of femur

Posterior cruciate ligam

Anterior cruciate ligame

Medial collateral ligame

Medial meniscus

Meniscal ligament

Gastrocnemius muscle (medial head)

ether, static (*non-contractile*) and dynamic (*contractile*) restraints make joint stability possible. The bones, cartilage, muscles, ligame tendons, capsule, and other soft tissues all work together to provide stability.

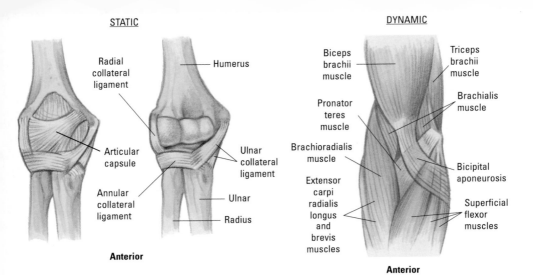

Anterior

Anterior

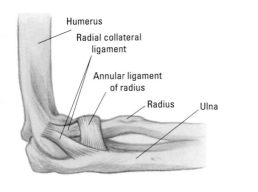

Lateral

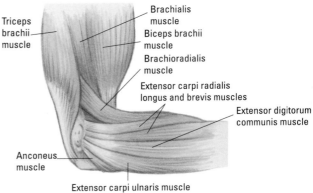

Lateral

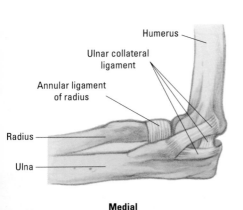

Medial

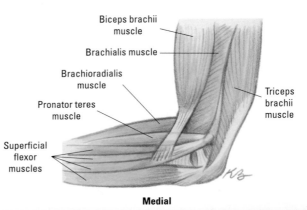

Medial

Stability is vital to maintaining good health and avoiding injury. The dynamic and passive restraint mechanisms help keep your

Fibrous and Cartilaginous Joints

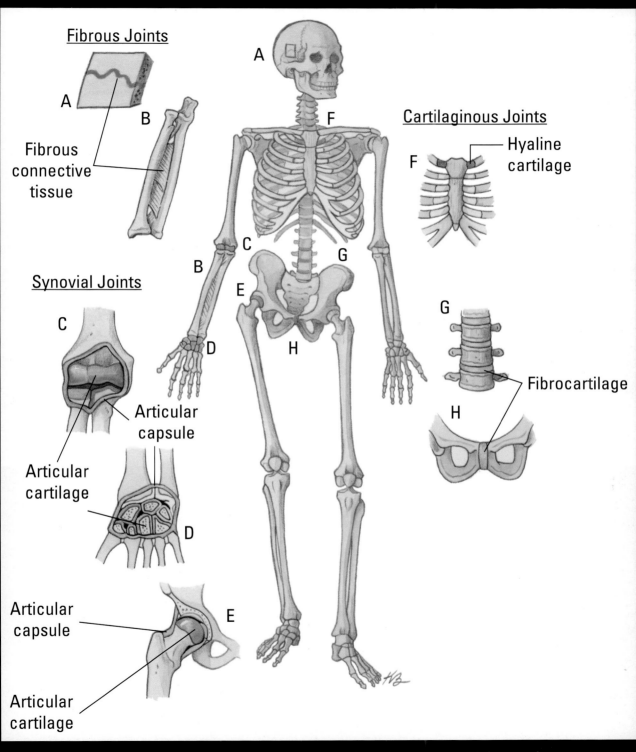

Fibrous Joints

A

Fibrous connective tissue

B

A

Cartilaginous Joints

Hyaline cartilage

F

F

B

C

G

E

Synovial Joints

C

Articular capsule

Articular cartilage

G

Fibrocartilage

D

H

H

Articular capsule

Articular cartilage

E

Joints often serve a unique purpose. The different types of joints depicted here are responsible for unique functions and levels of stability.

The Sarcomere

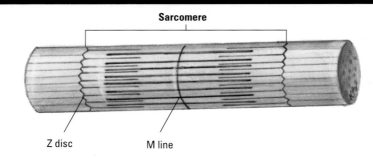

Sarcomere

Z disc

M line

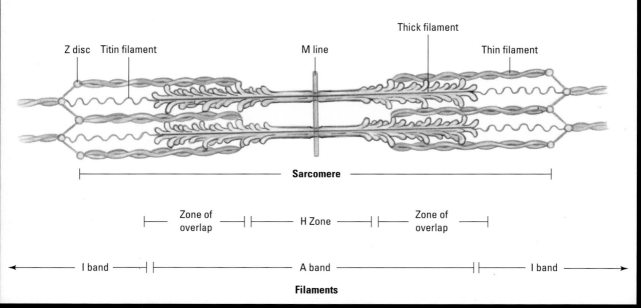

Z disc Titin filament M line Thick filament Thin filament

Sarcomere

Zone of overlap H Zone Zone of overlap

I band A band I band

Filaments

Muscles are made up of many sarcomeres connected into fibrils and wrapped in bundles. All force for muscle contraction happens when cross bridges connect and rotate in the sarcomere.

The Motor Unit

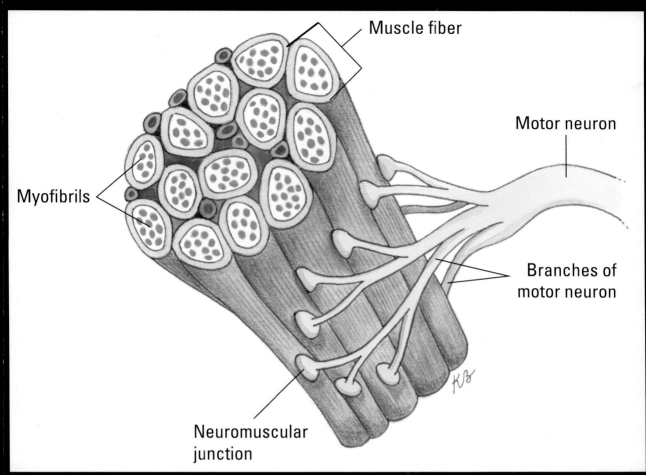

Muscle fiber

Motor neuron

Myofibrils

Branches of motor neuron

Neuromuscular junction

A muscle fiber will not contract unless the motor nerve that attaches to it is stimulated. There are many motor units. By controlling how many motor nerves are stimulated, you can control the force you generate.

Muscle Pennation

Fusiform

Pennate

Bipennate

Flat

Triangular

Strap

Multipennate

Some muscles contribute to significant amounts of motion; others barely cause any movement at all. A muscle's construction is a primary determinant of the kind of movement that's possible. Here, you can see the various types of muscles and their fiber arrangements.

Agonist/Antagonist

Humerus bone

Concentric/Agonist:
Biceps muscle contracting

FLEXION

Eccentric/Antagonist:
Triceps muscle
relaxing

Eccentric/Agonist:
Biceps muscle relaxing

Concentric/
Antagonist:
Triceps muscle
contracting

EXTENSION
Lowering
against
gravity

K. Born

The agonist muscle works to accomplish the desired movement. In this case, it acts both eccentrically and concentrically,
depending on the intent of the movement.

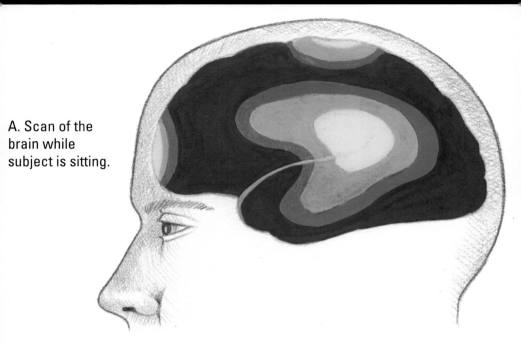

A. Scan of the brain while subject is sitting.

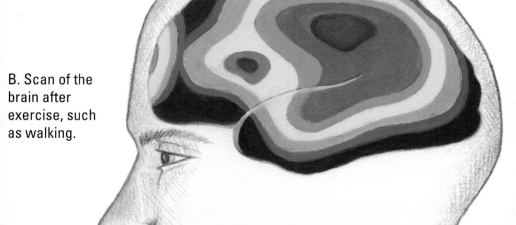

B. Scan of the brain after exercise, such as walking.

Exercise or any physical activity increases blood flow in the body. Added blood flow to the brain means the brain is

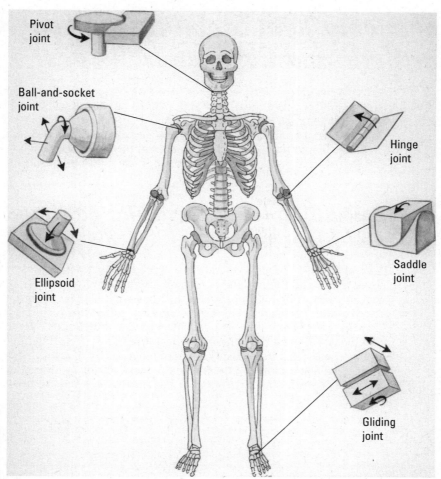

Figure 9-1:
Types of
joints.

Illustration by Kathryn Born, MA

By degrees of freedom: Uniaxial, biaxial, and so on

Another way joints can be classified is by the number of axes of rotation that they allow, often referred to as *degrees of freedom*. Those joints that allow one, two, or three axes of rotation are referred to as *uniaxial, biaxial,* or *triaxial,* respectively. A biaxial joint has two degrees of freedom: an ellipsoidal joint (knuckle), for example, can produce flexion and extension (bending) as well as abduction and adduction (splaying). (Refer to Chapter 7 for a complete discussion of the axes of rotation and the different types of movement.)

Enhancing Joint Stability and Longevity: Cartilage and Connective Tissues

Although the joints between bones fit together nicely in many cases, a nice fit isn't enough to make the joint stable. What's needed is some type of connective tissue that can stabilize the joint and enable it to perform its given function. The most common connective tissues are tendons, ligaments, and joint capsules. In addition, joints often possess a protective covering — articular cartilage — that reduces friction and helps the joint move smoothly.

Smoothing it out: Articular cartilage and fibrocartilage

Several of the joints described in the preceding sections possess a protective covering around them. Called *articular cartilage* (or *hyaline cartilage*), this covering reduces friction between the bones and allows for smooth motion. The articular cartilage is a dense, white connective tissue that is typically very thin (see Figure 9-2). Its purpose is to distribute loads over a wider area, effectively reducing the contact between the two articulating bones, and to reduce the friction that exists during movement. By distributing loads over a larger area and reducing the friction present during motion, the articular cartilage protects the parts of the bones that are needed for movement.

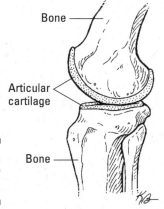

Bone

Articular cartilage

Bone

Figure 9-2: Articular cartilage.

Illustration by Kathryn Born, MA

Another kind of cartilage that preserves and facilitates movement in the various joints is articular fibrocartilage. *Articular fibrocartilage* typically takes the form of a cartilaginous disk (referred to as *menisci*, or *intervertebral disks*) between the bones. These fibrocartilagenous structures help with the following:

✔ Distributing loads over a joint's surface

✔ Improving the fit of articulating surfaces

✔ Limiting bone slip within a joint

✔ Protecting the edges where articulating bones touch

✔ Lubricating the articulating surfaces of bone

✔ Acting as shock absorbers

Think of articular cartilage as more of a sheet covering the surface of a bone. Articular fibrocartilage, on the other hand, is typically a piece of cartilage located *within* a joint; it provides a bit more cushion during function.

The articular cartilaginous structures are vital to maintaining the health of the joint throughout a person's lifespan. Yet the cartilage can be damaged when large amounts of force are delivered or when repetitive stresses are applied to the area over time.

Holding it all together: Articular connective tissue

Although the joints fit together nicely in many cases, the bony connection alone wouldn't provide enough stability to enable you to perform the many activities that you commonly engage in. Fortunately, you don't have to rely solely on how well your joints fit together because connective tissues — tendons, ligaments, and joint capsules — provide the extra support required for movement:

✔ **Tendon:** A *tendon* is a soft tissue structure that connects a muscle to bone. Often the tendon attaches the muscle to a movable aspect of the joint. The amount of movement created by the tendon depends on the size and length of the tendon, along with the type of joint it attaches to. A tendon can be overstressed, a condition referred to as *tendonitis*, the swelling of the tendon.

✔ **Ligament:** *Ligaments* connect bones to other bones to keep them organized and in their proper place. The wrist, for example, has a number of gliding bones. The ligaments in the wrist keep these bones from becoming disassociated and unorganized, which can result in injury and lack of use. An injured ligament is referred to as a *sprain*. Commonly people sprain the ligaments in their ankles and knees.

✔ **Joint capsules:** In addition to the tendons and ligaments within or around a joint, a joint capsule may also exist. Joint capsules are made up of dense, soft tissues that provide stability and facilitate the function of other structures. For example, a joint capsule provides a connection among ligaments, tendons, and articular fibrocartilage, while also assisting in creating a capsule in which synovial fluid can freely assist in decreasing friction during movement.

These soft tissue structures are considered *non-contractile,* meaning they're static, providing support and assistance with movement but not exerting force themselves, which is the muscles' job. When a muscle contracts, its force is delivered to the affected bone through the muscle tendon. Like tendons, ligaments also absorb and/or deliver forces to various portions of the joint.

Tendons and ligaments are *extensible;* that is, they're able to stretch when force is applied. When the force is relatively low, they can return to their normal length; however, when the force is large enough to stretch these structures beyond their elastic limit, they become damaged and cannot return to their normal length without surgery. When tendons and ligaments are stretched too often and/or too far, they become loose, and their function is compromised.

Getting Physical: Understanding the Functional Basis to Moving

For your body to move the way it's supposed to, not only do the individual structures — muscles, tendons, cartilage, and bones — have to function as they're supposed to, but they also must work in an interrelated way to provide stability and normal function. This section investigates the mechanisms responsible for providing stability and explains how such a complex feat as coordinated movement is accomplished.

Perusing the factors that affect stability

Have you ever sprained your ankle, limped around on bad knees, or felt your shoulder pop out? These events all describe what can happen if your joints lack stability. When joints are stable, they allow the bones to articulate, or move, without a lot of displacement. In other words, the bones move the way they're supposed to; they don't "pop" out of place.

Each joint possesses a unique requirement for stability. As we explain in the earlier section "Functional classifications: Synarthroses, diarthroses, and more," your body contains immovable, slightly movable, and freely movable joints. Factors that play a role in stability involve the bony and soft tissue structures that support the joint: how the bone is shaped, the arrangement of the ligaments, and more.

The shape (and contact points) of things to come

One of the factors that most influences stability is the shape of the articulating bones. Typically, bones that make up a joint are shaped as opposites to their counterparts. Where one bone ends in a socket, for example, its counterpart will end in a "ball" (refer to the earlier section "Freely movable joints: Diarthroses" for a discussion of the different kinds of articulating joints). This arrangement is often referred to as the *convex and concave orientation,* and it allows for increased stability.

Joints also tend to be more stable at certain points in their range of motion:

- ✓ **Closed-packed position:** When the articulating surfaces of a joint are in a position where the most amount of each bone is in contact with the other, they are said to be in the *closed-packed* position. In this position, stability is increased.

- ✓ **Loose-packed position:** When the articulating bone surfaces are in less than maximum contact, the joint is in a *loose-packed* position (also referred to as an *open-packed* position).

The more surface area of contact, the more stable a joint will be. The amount of surface area contact between joints varies from person to person. Bony differences between people and past injury to the bone or soft tissue support structures are potential causes of decreased stability.

It's articulation time: Do you know where your ligaments are?

Muscles, ligaments, and tendons all connect to the joint and provide for the delivery of or resistance to forces. The arrangement and integrity (condition or strength) of these structures play major roles in maintaining stability.

Ligaments, for example, attach to the bones and resist tension, thus helping to keep the bones together. When the bones are kept together within the joint, stability is enhanced. When muscles contract and exert forces on the bones via the tendon, the bones typically move closer to one another, maximizing stability.

How tight or loose are you?

Each of your joints relies heavily on the muscles around the joint to provide movement through contraction. When the muscles have adequate strength and length, the function is good. Yet a muscle imbalance around a joint — when one muscle exerts more force than the other — can actually lead to a destabilizing situation. If, for example, your knee extensors (quads) are a lot stronger than your knee flexors (hamstrings), when you contract your quads, they'll overpower your hamstrings and either injure the hamstrings or cause your joint to move beyond its normal range of motion.

The muscles around a joint should be strengthened together and in a functional way. Doing so ensures that they are all strengthened for that particular function and maintains the structural balance required for stability. So rather than just doing knee extension or flexion, for example, be sure to develop the muscles that are used for all the other motions that are involved with your activity.

Long or short? It matters

Another factor within the musculature that may impact stability is the length of the muscle. Revisiting the hamstring and quadriceps example introduced in the preceding section, if either of these muscles (or muscle groups) has limited flexibility, you won't be able to achieve the appropriate position that may be required for the activity you're attempting. If you have tight hamstrings, you may not be able to extend your knee far enough to achieve a normal heel strike while you walk, for example, a situation that has implications with the rest of the activity and, ultimately, the stability of the joint.

Most ligaments and tendons attach to the joint in a way that maximizes stability, and both adapt to the forces that are applied. Over time, they may *atrophy* (shrink) or *hypertrophy* (get bigger), depending on what is required of them. This situation increases the chances of injury or re-injury. If you have previously sprained your ankle, for example, chances are that that loose ligament will make you prone to another injury. Maybe you'll injure another structure, or maybe you'll sprain your ankle again.

The role of other connective tissues

In addition, to the ligament, tendons, and muscles, other connective tissues exist in and around your joints that impact stabilization. For instance, the joint capsule (explained in the section "Holding it all together: Articular connective tissue") and *fascia* (the connective tissue that surrounds and connects the muscles and other soft tissues) may play a role. These soft tissues themselves may help to stabilize the bones by either providing points of attachment for the tendon and/or ligaments or helping to facilitate sensory input from the joint and muscle activity.

Understanding restraint mechanisms

Joint stability isn't the job of any single structure doing its thing in isolation. Instead, it's the result of the muscles, tendons, ligaments, bones, and other soft tissues all working in a coordinated, interrelated way. All these tissues possess mechanoreceptors responsible for sensing and assessing movement; they all provide a unique attribute to stability, which is enhanced when each completes its specific task and communication is facilitated between the different components.

In this section, you explore two major restraint mechanisms — the active restraint mechanism and the passive restraint mechanism — that together maintain stability, enhance activity, and help you avoid injury.

To function properly and maintain stability, the active and passive restraint mechanisms need to work together. If problems exist in either mechanism — the muscles don't fire properly or if sensory information isn't collected and processed as it should be — stability is compromised. When *both* active and passive restraint mechanisms are on the glitch, the effects on stability are even more complex — and detrimental.

Muscling in: The active restraint mechanism

The *active restraint mechanism* can broadly be defined as the contractile component of joint stability — in other words, the muscles. Muscles are considered the *active* mechanism because they act on and around a particular set of structures.

Not only do the muscles that act on a joint need to provide the forces required to move or propel objects, but they also must resist forces that would otherwise cause injury. Think of a task that requires hand-eye coordination. When you reach out for and turn a doorknob, for example, all the muscles affecting your shoulder, elbow, wrist, and fingers must contract in a coordinated effort; if they don't, you won't be able to grasp the knob, twist it, and pull the door open.

To maintain stability while performing such a task, the muscles need to exert force in a coordinated fashion, do so in a timely manner, and be strong enough to generate the effect required:

 ✔ **Working in a coordinated way:** Your muscles provide coordinated function. Through motor patterns established over time (you've opened many doors in the past, for example), tasks (movement) that you perform are already strategized. Patterns have been developed to coordinate the task at hand while minimizing distracting or opposing forces. When the efforts of all the muscles involved in a given task are coordinated, stability is maintained, and you can complete the task more successfully.

✔ **Possessing the ability to receive and interpret sensory information in a timely manner:** Sensory information is continually being collected from the joints involved in an activity, and the forces needed to complete the task are determined based on this information (how fast you want to go, whether you're traveling up or down hill, whether the ground is even or uneven, and so on).

Depending on your ability to retrieve this information, your muscles adapt accordingly. If a muscle or group of muscles isn't able to interpret the information in a timely manner, or if the muscle or muscle group is weak, then the movement pattern is thrown off, and stability may be compromised. (The next section explains how these messages get to the muscles.)

✔ **Being able to react with adequate force:** Muscles that are too short or too long (essentially not in the ideal position) have difficulty getting the timing right or being able to develop the force necessary to either counter or facilitate the activity.

Simply contracting isn't enough. A muscle has to contract with enough force to complete its task. Many who have had previous injuries or who have neurological problems that affect strength have a hard time stabilizing their joints. When the communication network lets the joint's support structures communicate efficiently, function is facilitated because collaboration and timeliness are enhanced. The end result? Fewer injuries and less dysfunction.

How stimulating! The passive restraint mechanism

The structures that make up the passive restraint mechanism are those that are non-contractile and involve everything *except for* your muscles:

✔ **The cartilage, bone, ligaments, and other connective tissues:** Refer to the earlier section "Enhancing Joint Stability and Longevity: Cartilage and Connective Tissues" for details on these structures.

✔ **Mechanoreceptors:** *Mechanoreceptors* are specialized sensory receptors whose job is to detect neurological information.

Mechanoreceptors respond to mechanical stimuli such as tension (stretching), pressure (compression), and displacement (movement). They collect and help decipher the needed information regarding movement. For example, when pressure is applied to a bone, the mechanoreceptors transport that information to the brain to be evaluated and acted upon. Or if you've just stepped in a hole, mechanoreceptors ask certain muscles (ankle everters) to contract to minimize how much you twist your ankle. By trying to stop the ankle from twisting, injury may be avoided.

You can find mechanoreceptors in muscles, tendons, bones, ligaments, and other soft tissues. When these structures are stimulated or affected by an activity, the information is shared and one of two things happen: Either the information elicits a reflex response (you yank your hand away from the hot burner, for example), or it triggers the brain to create a new motor plan (sequence of activities like dodging a defender who is trying to block your shot).

Because mechanoreceptors need to be mechanically stimulated, their responses are susceptible to injured tissue. For example, a ligament that's been sprained in the past will be a bit longer. The next time you sprain your ankle, it will twist even further before the mechanoreceptor is activated and recognizes the situation.

ACL injury

Ligament injury is quite common in physically active people. One very common ligament injury involves the anterior cruciate ligament (ACL), in the knee. The ACL is responsible for keeping the tibia from sliding too far forward on the femur. Often the ACL gets injured when an athlete is decelerating or accelerating while simultaneously rotating, as you do when you stop and change direction quickly. ACL injury is so common in the sports world that many athletes either have or know someone who has had this injury. (Fortunately, an ACL injury isn't a career ending one; it can be successfully rehabilitated with time.)

Several factors may predispose athletes to ACL injury. One of the most common is a muscle imbalance that exists between the quadriceps and hamstring muscle groups. Typically, the hamstrings should be one-third weaker than the quadriceps. When hamstrings are more than one-third weaker, the predisposition for ACL injury exists. Another predisposing factor is the angle at which the knee is composed, called the *Q-angle*. A large Q-angle (called

knock knees) increases your chances of suffering from an ACL injury, as does a smaller space between the *femoral condyles* (the bony protrusions on either side of the bottom of the femur), known as the *intercondylar notch*. Two others thought to play a role are the hormonal changes that occur during menstruation and having flat feet, but as yet, the evidence connecting these factors and ACL injuries isn't as strong.

Unfortunately, females tend to possess more of these predisposing factors than men, and, not surprisingly, girls and women are affected more by this injury. Several preventative programs that integrate females into physical sport activity earlier have been shown to possibly curb the incidence of injury. These programs are aimed at strengthening the lower extremities. Some speculate that beginning physical activity earlier allows the neurological system to develop earlier and helps athletes avoid injury that otherwise may occur. In the end, your best bet is to emphasize good biomechanical form and strengthening.

Being flexible: You can do it!

Movement is the cornerstone of living a healthy and productive lifestyle. Whether you're getting up from a chair, walking down the stairs, throwing a baseball, or taking a jog, each task requires a certain range of motion (flexibility). You need to be familiar with two types of flexibility: static and dynamic.

Static flexibility

When most folks think about how flexible they are, they usually think about static flexibility. *Static flexibility* is the amount of joint motion exhibited when you move to the end of the range of motion and hold still — such as when you bend over to touch your toes, feel the intense stretch along the back of your legs, and then hold the position. This type of flexibility is usually very controlled and emphasizes only one joint or muscle group; in the case of touching your toes, it emphasizes the hamstring group.

A static assessment of how flexible you are is helpful in understanding the amount of total joint motion that a limb is capable of. When you examine a joint from the beginning to the end of its range of motion, you get an impression of the total amount of movement allowed, called the *envelope of motion,* within a particular joint. Typically, these assessments are performed when the person is relaxed. Makes sense, doesn't it? If you're trying to stretch, you relax so that you can reach the end of the movement.

Although some people seem to be more flexible than others, the amount of motion each limb and joint is capable of differs throughout the body. It's the integration of range of motion across all the joints that allows for the functional activities that you perform on a daily basis.

Dynamic flexibility

Because many joints must work in concert with others to accomplish movement, you can't simply look at static flexibility to get a holistic view of movement. Instead, you need to consider dynamic flexibility, which involves the interrelationships among joints.

Dynamic flexibility is the amount of motion experienced during movement and typically involves multiple joints simultaneously. Most activities that you participate in are *dynamic tasks,* meaning that the movements associated with them have multiple phases and parts. To throw a ball, for example, you need movement to occur at the elbow, shoulder, wrist, spine, and legs.

Unlike static flexibility, which typically involves a controlled stretch-and-hold type of movement, dynamic flexibility relies on the interrelated joint ranges of motion required to complete the task. If a limitation is noted in one of the involved limbs, a change will likely occur elsewhere. If your elbow doesn't rotate correctly when you throw a ball, for example, the ball won't go very far.

Assessing dynamic flexibility and recognizing the interrelationships that make everyday activities possible is of the utmost importance, especially for coaches and healthcare providers.

An assessment of dynamic flexibility takes into consideration the action of all the joints involved with a particular movement. For example, if you were assessing a person's dynamic flexibility by watching him walk, you would do the following:

- ✔ Note the discrete movements that make up the overall motion.
- ✔ Understand how each movement affects the other movements.

In this case, you may notice that the subject isn't able to fully strike on his heel when the foot contacts the ground. As a result, he is forced to make contact at the middle of his foot, the primary cause of which is that his knee didn't extend enough to get the heel down in time.

Although static flexibility is important and affects your ability to move with enough dynamic flexibility, the key component is how all the joints work together when a movement occurs. In the example related to walking, a static assessment may reveal that the subject had full range of motion when he tried to straighten his leg; the dynamic assessment reveals that, when the ankle, knee, and hip needed to move in an integrated fashion, a deficit existed that decreased the subject's range of motion.

Wrangling with range of motion

Within the musculoskeletal system, the *range of motion (ROM)* of a joint describes its flexibility (but we don't use the term *flexibility* in this context because that term is often used to describe how tight your muscles are, and it just confuses things).

Factors influencing joint flexibility

A number of different situations, events, and conditions can influence joint flexibility. The ROM of the joint is largely dictated by

- ✔ **The type of joint:** See the earlier section "Getting These Old Bones to Move: Types of Joints" for details on the different kinds of joints.

- ✔ **The shape of the bones forming the joint:** The hip joint, for example, is a ball-and-socket joint where the head of the femur inserts into the round cavity that is the acetabulum. Because the femur is firmly within the acetabulum, when the joint reaches extremes of flexion and extension, the bone just isn't able to move any further because, in doing so, it will contact the other articulating bone. Often clinicians refer to this stopping point as an *end feel,* and in this case, it would represent a *bony* end feel, which is abrupt, hard, and unforgiving. Another example would be elbow extension, when you simply reach the point at which your elbow can extend no further. (For more on end feel, head to the later section "Paying attention to how the end feels.")

✔ **The soft tissue structures that surround the joint and stretch:** If you can't reach over and touch your toes, chances are that your hamstring muscles are too tight and don't allow you to bend far enough. This is often referred to as the *tissue stretch,* and the end feel is pretty firm. Because of the inherent elasticity of the different soft tissue structures, some stretch, or give, more than others.

Try this example, which involves your fingers: Try bending them back toward your forearm. You probably notice that this motion is relatively easy to accomplish early on, but when you approach the end range, notice the resistance, which eventually stops the motion.

✔ **The soft tissue structures that block movement:** Soft tissues can impede the ROM simply because they get in the way. Called the *soft tissue approximation end feel,* just the squeezing of the tissue stops the motion. For example, you may not be able to bend your leg such that your heel is able to touch your bottom. In this case, the hamstring muscles are squeezed and don't allow for this motion to go so far. If you have large amounts of fat in this area, you'll have deficits in your ROM.

✔ **The articular cartilage, ligaments, joint capsule, and fascia throughout the body:** These structures are constantly changing as they respond to use and disuse. Typically, if you assume a position for long periods of time or if you habitually perform a particular movement, the tissues adapt. Such adaptations to movement can have negative effects on the health of the joint and surrounding soft tissues.

✔ **The relative elasticity, extensibility, and laxity that exist within the soft tissues around the joint:** A joint that is composed of tight muscles, ligaments, or tendons tends to be tighter and has less ROM than a joint whose structures are looser. Conversely, if these structures have been injured and are more lax, the joint may experience too much motion.

Although you may think that you can't be too loose, you absolutely can. When your joints are lax and allow more than normal ROM, added stress is distributed to the cartilage and bony structures. Sometimes, if the stresses are applied repetitively and at large enough levels, damage can occur.

Measuring ROM

The ROM within a joint is based on how far you can stretch and reach. Although this method is a fine way of assessing your motion, it lacks an objective measure to compare your results to. Sometimes, you need to know the actual amount of movement that exists within the joint(s) related to both dynamic and static flexibility. For example, after an injury, you can use joint measurement to assess how well you're healing. Often after surgery, the motion in your limb is limited, and it's important that you regain what you lost. Being able to gauge your progress and know how much movement is lacking is a key part of rehabilitation.

Joint ROM is measured in units of degrees. The reference point for most measurements is the body in anatomical position. When a joint is in anatomical position (refer to Chapter 7), the joint is said to be at *zero degrees,* or *neutral.* From this point, you can measure the amount of motion that exists. The ROM for flexion of the elbow, as shown in Figure 9-3, for example, is considered to be the angle created when your limb moves from the fully extended (anatomical) position to the maximal point of flexion. To figure the amount of extension within the joint, you simply return the limb from the maximal flexion to its anatomical position.

Measuring ROM

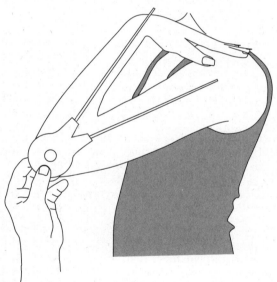

Figure 9-3:
Measuring
range of
motion.

Illustration by Wiley, Composition Services Graphics

You can measure ROM with these tools:

✔ **A goniometer:** Any joint motion is able to be quantified using goniometry. *Goniometry* is the act of measuring angles, in this case, joint angles. Goniometry is usually measured using a *goniometer,* a device that, if used correctly, can easily quantify joint motion. Goniometers come in different sizes and allow for either 360° or 180° of motion.

✔ **An inclinometer:** Inclinometers can be purchased from your local hardware store and strapped onto the limb; they simply tell you how much movement the limb has gone through. The principles for using this device are similar to those used with a goniometer.

Paying attention to how the end feels

Each joint has an anticipated normal ROM. What constitutes "normal" is based on studies of thousands of people. When people fall beyond these norms — that is, they either have greater or lesser ROM — they experience effects within the affected joint and beyond it, a situation directly related to the idea of dynamic flexibility (explained in the earlier section "Dynamic flexibility").

Additionally, due to the anatomic makeup of each joint, a normal end feel is established. *End feel* describes how a joint comes to the terminal portion of its movement. The end feels are *bone to bone, soft tissue approximation,* and *tissue stretch,* as we describe in the following list:

✔ **Bone-to-bone end feel:** A *bone-to-bone end feel* exists when the joint's ROM is limited due to bones contacting one another. This end feel is abrupt, hard, and unforgiving. A very common example of this is an elbow extension. When the elbow extends, the hook-shaped protuberance from the ulna (called the *olecranon process*) contacts a depression on the humerus (called the *olecranon foss),* and the motion is stopped.

✔ **Soft-tissue–approximation:** A *soft-tissue–approximation end feel* involves a compression of the soft tissues around a joint that leads to the end of the motion. An example of this is elbow flexion, when the lower arm is flexed toward the shoulder. This motion is usually terminated because the biceps muscles are compressed in a way that makes further motion impossible.

✔ **Tissue stretch:** *Tissue stretch* refers to a hard or firm end feel with a slight give to it. The inherent elasticity of the different soft tissue structures cause some to stretch, or give, more than others, dictating the level of end feel produced as tissue stretch. Typically, this type of end feel involves progressive tension that continues until motion is stopped, as happens, for example, when you bend your fingers back toward your forearm. Initially, the motion is relatively easy to accomplish, but as you approach the end range, resistance eventually stops the movement entirely.

You need to understand the joint's underlying anatomical makeup to know what to expect as the normal end feel for that joint and movement. If you experience an end feel other than what you expect, an injury has probably occurred. Evaluating end feel is how clinicians can tell whether you've torn a ligament, for instance. If your bone moves beyond what is normally expected, the clinician may conclude that you've damaged your ligament, which is meant to keep the joint stable.

Being lax isn't necessarily a good thing!

Don't confuse being flexible with being lax. Although sometimes related, the two terms refer to slightly different concepts. *Flexibility* often refers to the range of motion of a joint, whereas *laxity* refers more directly to the movement within a particular joint. For example, being able to glide, or *translate,* your shoulder to a point where it is close to popping out is an example of laxity; the ability to reach behind you and scratch your back is more a measure of flexibility.

Laxity within a joint can be a good thing, but it can also be detrimental, negatively impacting the stability within the joint and potentially contributing to injury. Most of your joints allow for some amount of translation between the bones; many times, this translation allows the joints to move in the direction(s) you need

them to. Yet when the laxity within the joint allows for too much translation, it can result in abnormal movement and an unstable environment. For example, if a lax shoulder rotates too far back into external rotation, it will also translate, requiring the rotation to occur at the edge of the joint. When this happens, the joint is asked to perform in a position that it wasn't made to accommodate. In situations like these, the function causes undo stress, and the joint and/or articular cartilage may be damaged. In extreme cases of too much laxity, the result may be a dislocation.

To counter the effects of laxity on a joint, you need to strengthen the musculature surrounding the joint. When you strengthen the joint, you improve how stable the joint is during activity.

You Want Me to Put My What Where? Stretching Redefined!

Stretching has long been thought of as a necessity to physical activity. Before nearly every soccer or baseball practice, your coach made you stretch out. Today, however, the effectiveness of stretching routines is being questioned, and much debate exists about the effectiveness and type of stretching that you should do, if any (if your practices were anything like ours, you spent more time catching up on the gossip of the day than actually stretching). This section explores the physiological components of stretching and answers the questions related to the types of stretching you should or shouldn't do.

Looking at what happens when you stretch

Increasing or maintaining motion in a joint requires that you use and stretch the joint on a regular basis. Why? Because the soft tissues adapt to the stresses that are applied to them, whether those stresses are caused by sedentary lifestyles, habitual positioning, and/or repetitive activities.

In the sedentary individual, the soft tissues are likely to tighten and result in less joint ROM. For those who sit in a certain position for extended periods of time or who are active but complete the same task the same way over and over again, the body may produce an increased ROM in certain parts of the movement while tightening up and exhibiting less motion in others. Additionally, a decrease in ROM in a major joint may actually cause decreases in other related movements. For example, not using your thumb often leads to decreases in the motion in the wrist.

These effects occur because of the mechanisms that come into play when you move or, as the case may be, don't move. In this section, we explain these mechanisms in detail.

Making the stretch possible: Autogenic inhibition

All your muscles contain organs that are able to detect various types of sensory information. These structures are called *mechanoreceptors*. When stimulated, they communicate with the central nervous system (CNS) and tell it what's happening within that structure. The stretch reflex carries the most responsibility when stretching and relies on two mechanoreceptors to guide its function: the muscle spindle and Golgi tendon organs (GTOs), which are responsible for detecting changes in muscle length and tension (take a look at Chapter 10 for more on these structures).

When the muscle is stretched, both mechanoreceptors begin to send signals to the spinal cord and brain. Here's what happens:

1. **At first, the muscle spindles react by contracting the muscles to prevent too much stretching in that muscle.**

   Obviously, if you're trying to stretch and the body responds by minimizing the stretch, it won't be very productive.

2. **If the stretch is held for a few seconds, typically ten seconds or so, the GTO impulse overcomes the muscle spindle response.**

   The job of the GTOs is to share the information about level of tension and length of the structure to the spinal cord.

   Unlike the contraction that's initiated when the muscle spindles are activated (Step 1), the GTOs encourage a reflex relaxation of the muscle, which helps to prevent injury. By decreasing the tension within the muscle, the reflex keeps the muscle from extending beyond its limits. When the stretching muscle relaxes as a result of the GTOs within it, *autogenic inhibition* is achieved. This mechanism makes stretching possible.

Opposing muscle groups working together: Reciprocal inhibition

Activity is often dependent on opposing muscles around a joint working in concert. For example, as your quadriceps contracts, your hamstring needs to relax a bit to allow the knee to move. If both muscle groups continue to contract, they end up fighting each other, making movement inefficient at best.

When a muscle group is stimulated, it may actually have a relaxation effect on the opposing muscle, a phenomenon referred to as *reciprocal inhibition*. In reciprocal inhibition, one muscle group (the *agonist*) is contracted, and the opposing muscle group (the *antagonist*) is relaxed. The GTOs of the opposing muscle are stimulated and decrease the tension in that muscle, allowing for more efficient motion because the agonist doesn't have to work as hard to achieve movement. You can use both of these techniques to your advantage when implementing stretching routines. Figure 9-4 shows an example of reciprocal inhibition.

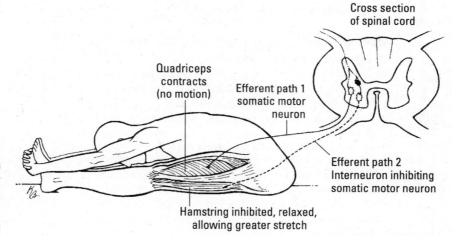

Figure 9-4:
Reciprocal
inhibition.

Cross section
of spinal cord

Quadriceps
contracts
(no motion)

Efferent path 1
somatic motor
neuron

Efferent path 2
Interneuron inhibiting
somatic motor neuron

Hamstring inhibited, relaxed,
allowing greater stretch

Illustration by Kathryn Born, MA

It's more than just the nerves: Collagen, elastin, actin, and myosin

Another mechanism that assists with increasing flexibility involves the muscle structure itself, which has both contractile and non-contractile components:

✔ **Non-contractile components:** The muscles and tendons are both composed of collagen and elastin fibers, which are non-contractile. *Collagen* allows the tissue to resist forces that lead to tissue manipulation, and *elastin,* which is composed of elastic tissues, helps the tissues return to their original shapes.

✔ **Contractile components:** These include *actin* and *myosin,* which are the myofilaments responsible for muscle contraction. (For more on the muscles, check out Chapter 10.)

Altogether, these portions of the tendons and muscles work together to determine the structure's ability to be stretched and return to normal after being stressed.

The collagen and elastin along with the actin and myosin seem to resist the stretch that is applied to the structure; however, the force behind the stretch and the velocity with which that force is delivered determine how far the muscle stretches. During stretching, temporary changes in the length of the structure occur. Typically, these changes are short-lived, and the muscle returns to normal.

Permanently lengthening a muscle or tendon is difficult to do safely, but it can occur when elongation exists in response to long periods of stretching. Although the changes that occur in the muscle's physical and mechanical structures are usually temporary, continued activity is considered beneficial if you want more permanent changes.

The Pushmi-Pullyu of stretching: The balancing effects of agonists and antagonists

When reading and studying about stretching, you'll come across the terms *agonist* and *antagonist,* which describe what and how the muscles work to create a stretch (we introduce these terms in the earlier section "Opposing muscle groups working together: Reciprocal inhibition").

To review, the *agonist* is the muscle that is contracted to create movement and cause a stretch. The *antagonist* is the muscle that responds to the action of the agonist by being stretched. When you flex your elbow, for example, the muscle creating flexion is the biceps, which is the agonist, and the muscle that stretches is the triceps (antagonist). Balance between the agonist and antagonist is very important; otherwise, injury can occur due to muscular imbalances.

Stretching techniques

Maintaining full, unrestricted ROM is essential to your normal daily activities. The loss of flexibility in a joint can result in uncoordinated and awkward movement patterns. Although, without question, flexibility is essential to normal everyday activities and physical performance appears to be best when participants are loose, researchers disagree significantly on the effect stretching may have on performance and injury. Still, everyone agrees that joints need a certain amount of physiological ROM within them; without it, the body is affected.

The goal of all stretching routines should be to increase the ROM within the joint, and increasing ROM is a common rehabilitative exercise for both those who have been injured and athletes. In the following sections, we describe a variety of techniques that can enhance the flexibility of a particular joint. Some of these techniques are more specialized than others, and some require the help of someone else.

All the stretching techniques we list can help you maintain and, if done routinely, increase ROM. However, not all types of stretching are meant for everyone. Make sure you understand the risks associated with each technique and progress appropriately through your program. When in question, consult an appropriate professional to prescribe and guide your program.

Active and passive stretching

All stretching happens either actively or passively, and some stretching techniques use both:

- ✔ **Active stretching:** In *active stretching,* you contract the agonist muscle, which creates a movement in the joint that results in the antagonist muscle stretching.

- ✔ **Passive stretching:** In *passive stretching,* you rely on gravity or the force of another person to move the joint through its ROM to the point where a stretch is felt.

Both techniques stretch the involved tissues, and because of the different functions of and structures in the body, both techniques are needed. Early in the rehabilitation program following knee surgery, for example, the best strategy may be for you to extend and flex your knee on your own (active stretching). As you progress and the pain becomes less, the clinician typically begins to push on the limb as a way to increase the range of motion in a slow and controlled, but forceful, manner (passive stretching — which is not like a torture chamber, although that may be what you're thinking!).

Static stretching

The most common stretching technique is the static stretch. In *static stretching,* you rely on gravity, your body weight, or another person to create the push that results in the stretch. The chosen antagonist muscle is pushed to a point of maximal stretch and then held in that position for a period of time. Some people say that a short (five second) stretch is enough; others hold the stretched position for as long as a minute.

Ballistic stretching

In *ballistic stretching,* repetitive bouncing movements create an increase in ROM. This technique has been used for many years but has not been without its critics, who are concerned that the ROM created during the bouncing actually causes injury to the involved tissues, which are forced to move beyond what they may be able to handle safely.

Despite the controversy, ballistic stretching is widely used and has been shown to increase ROM. When engaging in ballistic stretching, make sure that the intensity of the bouncing doesn't create undo stress on your joints, which can result in muscle or joint injury.

Dynamic stretching

Dynamic stretching is an active and controlled technique that mimics functional activity. In other words, it includes movement patterns that you actually use in the activity you're getting ready to engage in.

Dynamic stretching has grown in popularity over the past several years, especially in the sporting world. Most sporting activities require repetitive motion of a particular task — throwing a baseball in practice, for example. After throwing the ball 50 or so times during a game, for example, players often experience soreness or tightness in the muscles of their throwing arms. The active warm up not only helps players loosen up and gain flexibility, but by getting them warmer, it also has long been considered an important aspect of getting ready to participate.

Proprioceptive neuromuscular facilitation (PNF): An advanced stretching technique

Proprioceptive neuromuscular facilitation (PNF) is a series of stretching techniques that involve particular combinations of contraction and relaxation of the agonist and antagonist muscles. We explain the more common types of PNF in this section. All are typically done with multiple (three or so) repetitions, and each phase is held for approximately ten seconds. The timing is important because it maximizes the GTO response, which is one of the primary factors in increasing flexibility in this technique (refer to the earlier section "Making the stretch possible: Autogenic inhibition").

When using the PNF technique, heed this advice to avoid injury:

- **Make sure the partner doesn't push too hard.** Remember, the technique *starts* at a point of maximal stretch, and further movement beyond that is the goal.

- **Don't engage too quickly when you go from pushing to relaxing and vice versa.** As the name indicates, the transitions and contractions should occur slowly.

This technique involves successive bouts of stretching and results in significant increases in ROM.

Hold-relax PNF technique

The hold-relax PNF technique involves an *isometric* (no movement) contraction of the hamstrings (antagonist) at the point of stretch. After pushing against resistance for approximately ten seconds, the leg is then pushed to create further stretch. After the ten seconds of pushing, the patient can actually be pushed further — evidence of the antagonist muscles relaxing and allowing more movement. Although other PNF techniques may require the use of a partner, you can actually complete the hold-relax technique by yourself.

Contract-relax PNF technique

The *contract-relax PNF* technique involves an *isotonic* (movement) contraction of the hamstring (antagonist) muscles. Here, the stretch begins at the point where a stretch is felt (the leg is lifted up in the position of a hamstring stretch); after the patient is at the point where he feels a stretch, he's asked to actively push his leg against the clinician and then to return his leg to the ground. After the leg is back on the ground, the clinician again lifts the leg while the patient pushes his leg into a stretch, most likely moving past where the last stretch occurred.

Slow-reversal-hold-relax PNF technique

The *slow-reversal-hold-relax* technique, shown in Figure 9-5, involves a little more coordination and communication between the player being stretched and her partner, as these steps show (we use the hamstring stretch here, but the process is generally the same when using this technique to stretch other muscles):

1. **The partner moves the player's knee and hip back to a point where she experiences a stretch.**

   Usually felt behind the knee and at the back of the leg, this is where the antagonist is.

2. **Upon feeling the point of stretch, the player actively attempts to straighten her leg back out and push down toward the ground, but the partner doesn't allow the leg to move.**

   This goes on for about ten seconds.

3. **The player contracts her agonist (quadriceps) and pulls her leg actively back toward her head to intensify the stretch. While the player actively pulls her leg back, the partner applies a slight overpressure and helps the player continue to move back.**

   Working together, the partner and player are able to achieve a maximal stretch of the antagonist muscle in that direction.

4. **When a new point of hip flexion and an increased stretch has been achieved, the player relaxes while the partner holds the stretch for about ten seconds or so.**

5. **After the rest period, the player repeats the preceding steps from the new position.**

1. Passive stretch

2. Contract against resistance

Figure 9-5:
The PNF
stretching
technique.

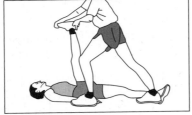
3. Relax passive stretch

Illustration by Wiley, Computer Services Graphics

Alternative stretching techniques

The stretching techniques discussed in this chapter are the more common ways you can increase ROM and flexibility in your muscles and joints. However, plenty of other common activities can result in the same increases in flexibility. Many people, for example, regularly engage in Pilates and yoga activities. These exercise and overall health and wellness programs incorporate similar strategies that are aimed at increasing both strength and flexibility.

Another technique that is gaining a lot of popularity, especially with the runners, is *soft tissue mobilization*. Although not necessarily a stretching technique, soft tissue mobilization can increase ROM of a joint and create added flexibility in a muscle. It involves self-massage (usually via large foam rollers, but anything that can provide pressure in an area will do the job, like a softball, your elbow or thumb, and so on) and muscle relaxation. By enhancing blood flow and the GTO response for muscle relaxation, this technique helps the affected muscle loosen up so that it can move better and with less discomfort. Most people use this technique at the gluteal (buttocks), hamstring (back of upper leg), and IT band areas (outside of upper leg), but any muscle group can benefit.

Other things to keep in mind about stretches and stretching

The benefits of stretching go beyond just what happens in the muscle; they also extend to the involved tendons and joint capsules. Here are some other things to know about stretching and stretching regimens:

- ✔ **Although an increase in ROM is likely the goal of a program, it is not always needed to get results you seek.** In many cases, just simply encouraging and facilitating the muscles to relax or loosen up helps the joints that they attach to. Many runners, for example, experience *runner's knee,* in which, over time, the runner experiences pain on the outside of his knee. The culprit is typically a tight illiotibial (IT) band. In treating this injury, increasing the knee's ROM itself doesn't eliminate the pain. Instead, the clinician loosens up the particular muscle so that it can go through the full ROM required by the activity without absorbing too much stress or having to overwork.

- ✔ **You can often achieve desired results by focusing on areas other than the problematic joint.** My (Brian) patients often say, "You know that I have discomfort in my knee, so why are you spending so much time stretching out my hip?" The reason is that everything in the body is integrated; the knee (in this example) is reliant on the other limbs to function properly. Because your muscles often cross multiple joints, they can be affected by various components at several different places. Knowing and understanding how your anatomy works is key to establishing a good stretching routine. The next time you have knee pain, try to stretch out your hip and see how it makes your knee feel!

Perusing Common Joint Injuries

Throughout the course of your life, you will inevitably experience some sort of injury to your joints. Some will be mildly debilitating; others may require significant rehabilitation and possibly surgery. Following are some of the more common issues that you may find yourself dealing with:

- ✔ **Sprains:** A *sprain* involves the disruption of a ligament within a joint. A ligament connects bony structures to other bony structures (refer to the earlier section "Holding it all together: Articular connective tissue"). Most sprains are not extraordinarily debilitating and only require a few days to recover. However, in some cases, the injury can be severe enough to warrant surgery, either to repair the ligament itself or to repair other structures that were damaged as a result of the sprain.

✔ **Dislocations:** A *dislocation* occurs when a joint becoming separated; that is, the two articulating bones come apart, leading to deformity and lack of function. Fingers are commonly dislocated at the joint when baseball players slide head first into the base. Another joint commonly dislocated is the shoulder joint (the glenohumeral joint) where it "pops out" and hurts like the dickens.

A dislocation is different from a fracture, even though they often look similar. For that reason, never attempt to *reduce* (pop back in) a dislocation. Without x-rays, you just never know whether you're dealing with a pure dislocation or a fracture that, if moved, may cause significant damage.

✔ **Bursitis:** *Bursas* provide lubrication and padding to many joints. Think of them as little oil cans that secrete synovial fluid into the joint when squeezed. The synovial fluid lubricates the joint and its structures. Many folks, especially older folks, experience bursitis. Bursitis commonly occurs in the elbows and knees; you may feel discomfort when you move and notice what looks like a rounded squishy growth protruding from the joint. In most cases, bursitis is caused from overuse or not appropriately increasing the intensity of exercise.

✔ **Arthritis:** Inflammation and pain experienced within a joint is often diagnosed as arthritis. The most common type of arthritis is *osteoarthritis,* a degenerative joint disease caused by the damage to articular cartilage. When healthy, the bones that make up your joints appear white and shiny. As a protective measure, the articular cartilage absorbs forces, lubricates, and provides for better contact within the joint.

Over time or because of injury, this cartilage may become damaged and/or degenerate. When this happens, the hyaline cartilage (the cover coating the bones; refer to the earlier section "Structural classifications: Fibrous, cartilaginous, and synovial") has to absorb these forces and provide the function of the articular cartilage. Unfortunately, it's not made to do this and quickly begins to degenerate, resulting in pain, inflammation, and additionally bony changes. These changes continue to be aggravated with physical activity and often lead to the need for a joint replacement.

Good joint health is a key component of leading a healthy, long life. So for your joints and for your overall health, make this your mantra: "Move in all ways and always move!"

Chapter 10

Let's Move, Baby! The Muscles

*N*early every aspect of your life involves movement. Even as you sit and read this book, you're using your muscles: You're sitting in a chair or lying down, holding the book, turning the pages, moving your eyes across the page, and so on. Movement can be fairly simple, as in the reading example, or it can be quite complex. Not only does your body depend on your bone structure to support movement, but it also depends on the muscles. Without your muscles, you can't move! Nothing can happen.

The effect muscles have on movement is dictated by the foundations of the muscular makeup. How much force is generated, the amount of joint motion, joint stability, and muscle action differ across muscles, based on their structure. In this chapter, you explore the different types of muscles, how they are organized into different types of fibers, and how you can use these different fibers to run fast, for example, or run for long periods of time. The variety in your muscles helps you perform a wide range of movements.

The Foundations for Muscle Movement: The Science behind Contraction

If you want to turn on a light, you have to flip the switch to send electricity to the light bulb. If you want to "turn on" a muscle — that is, make the muscle contract — you need to flip a switch in the brain. This "switch" is an action potential that originates in the brain from the motor control centers.

As we explain in Chapter 3, various parts of the brain send signals down the spinal cord to initiate movement and cause the muscle to contract. These signals are one part of the story of movement. The actual mechanics of muscle contraction are the other part of the story. Any movement at a joint requires that the muscles connected on each side of the joint shorten.

"How is that even possible?" you ask. "And wouldn't an the entire muscle shortening produce *a lot* of force? What if you only wanted a little bit of force? How is that controlled?" Read on for the answers. As you'll see, many things influence how a muscle behaves.

A muscle contraction is a very complex piece of work. It involves nerves, different types of fibers, and different types of attachments, and it can span multiple joints. All these factors are significant contributors to the muscular system.

Uncovering the structure of the muscle

A muscle is really many muscles woven together. It's made up of many smaller units of muscle fibers, and each unit of muscle fibers is itself made up of components that, when all are working as they should, make movement possible. Figure 10-1 shows the structures of the muscle as you go to the smaller units. Refer to this image as you read the next few sections.

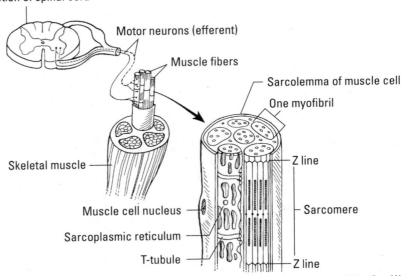

Figure 10-1:
Anatomy of skeletal muscle tissue.

Illustration by Kathryn Born, MA

Bundling up: Myofibrils

As Figure 10-1 shows, bundles of long fibers, called *myofibrils,* are grouped together to form a muscle. Each myofibril is made up of smaller, individual units of contracting tissue stacked end to end. The smallest unit that makes up the myofibril (and the one that does the contracting) is called a *sarcomere.* You can read more about the sarcomere and its role in contraction in the later section "The sarcomere and its parts: Shortening to produce force."

Releasing calcium: T-tubules and the sarcoplasmic reticulum

One ion that's very important for making a muscle contract is calcium. Although calcium is bound up in bone, it's also found in a system of storage vesicles, called the *sarcoplasmic reticulum,* within the muscle. Calcium can be released from this location and spread throughout a muscle via the T-tubules. All it needs is a stimulus, which it gets by way of the motor unit.

The motor unit: Connecting the nerve and the muscle

In the locations where the nerve actually reaches the muscle, the nerve doesn't just plug into the entire muscle. Instead, each motor nerve connects to only a certain number of muscle fibers. One nerve may connect to 100 fibers, for example, or it may connect to 1,000. This nerve-fiber connection is called a *motor unit.* These connections give your brain some control over just how many of those fibers contract.

The sarcomere and its parts: Shortening to produce force

As we note earlier, a myofibril is made up of a series of sarcomeres stacked end to end. Following are the key parts of a sarcomere (see Figure 10-2):

- **M line:** This is the connecting tissue located in the middle of the sarcomere, providing structure and stability to the sarcomere.

- **Thick filament:** Also know as *myosin,* it is pretty thick.

- **Thin filament:** Also known as *actin,* this protein filament is much thinner than myosin.

- **I band:** This band appears lighter in the sarcomere because it contains only the thin actin filaments. No myosin overlaps it.

- **A band:** This darker band contains an entire myosin (thick) filament.

- **Zone of overlap:** The action happens here! In this zone, actin and myosin connect and cause shortening of the sarcomere.

- **Z line:** Each sarcomere is connected to another sarcomere by rigid connective tissue called *Z lines.* The Z lines are essentially anchors that connect protein fibers. They provide stability to the sarcomere, and the pulling of myosin on actin moves the Z lines closer together during muscle contraction.

✔ **Actin:** Connected to the Z lines are thin protein filaments called *actin*. Actin has a twisted appearance, similar to what you'd get if you twisted a pearl necklace. It's a fiber that, when activated by a strong pull, can pull the Z lines closer together.

✔ **Myosin filaments:** Between the actin are thicker looking sets of protein filaments that have small "arms" coming out from them. These thick filaments, called *myosin filaments,* do the pulling. Think of the myosin as a rowboat in the water, and the arms as the paddles that do the pulling.

✔ **Titin:** Holding myosin in place and keeping it connected to everything is a large, springlike protein called *titin*. Titin, working a bit like a rubber band, helps give muscle an elastic property.

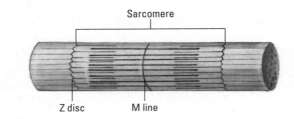

Sarcomere

Z disc   M line

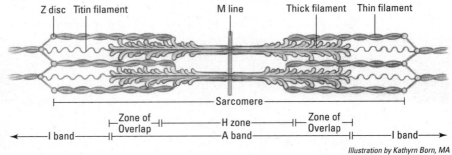

Z disc   Titin filament   M line   Thick filament   Thin filament

Sarcomere

Zone of Overlap   H zone   Zone of Overlap

I band   A band   I band

Figure 10-2: The sarcomere.

Illustration by Kathyrn Born, MA

Binding sites for muscle contraction

During a contraction, the sarcomere has to shorten, which happens when the myosin heads grab onto the actin at the binding sites and give them a pull. When a muscle is a rest (that is, not contracting), it can't grab onto the binding sites because the sites are covered. Uncovering the sites involves two proteins:

✔ **Tropomyosin:***Tropomyosin* is a long filament protein that lays on top of the binding sites on the actin. When tropomyosin covers the binding sites, the muscle is at rest.

✔ **Troponin:***Troponin* is a globlike protein that actually moves tropomyosin out of the way, if it has the right incentive, so that the binding site is exposed, a process we outline in the next section.

Filaments sliding past each other: Producing muscular force

Because you can't really see the motions within the sarcomere as it shortens, researchers have come up with a theory that explains what happens when the muscle contracts. Here is the step-by-step sequence of what makes the sarcomere shorten:

1. **The brain sends an electrical stimulus down the spinal cord and out to the motor units.**

2. **The motor units spread the signal to the fibers they're connected to, activating the release of calcium from the sarcoplasmic reticulum.**

3. **The calcium binds to the protein troponin, causing troponin to change its shape.**

4. **Troponin's shape change moves the tropomyosin out of the way so that the binding sites become available.**

    Figure 10-3 shows this sequence of tropomyosin movement and calcium binding.

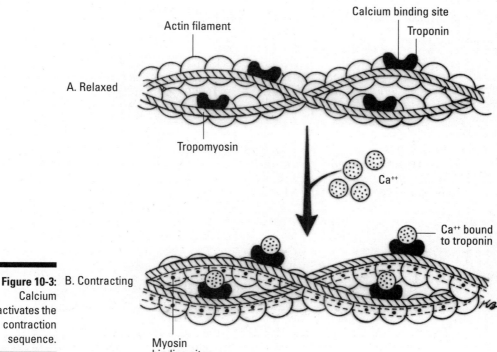

Figure 10-3: Calcium activates the contraction sequence.

Calcium binding site
Troponin
Actin filament

A. Relaxed

Tropomyosin

Ca⁺⁺

Ca⁺⁺ bound to troponin

B. Contracting

Myosin binding sites

Illustration by Kathryn Born, MA

5. **The myosin heads attach to sites on the actin, connecting the filaments.**

   In an earlier analogy, we say that the myosin is like a rowboat and the heads are the boat's oars. This step is akin to the boat putting its oars in the water.

6. **The myosin heads rotate and pull the actin on both sides of the sarcomere toward the center.**

   This action shortens the sarcomere and produces force.

7. **As long as there is energy to power the process and stimuli to keep it going, the myosin heads continue to rotate, releasing, grabbing, and pulling the next site and the next.**

   At this point, the oars of the rowboat are in full swing.

To visualize this process, put both your hands in front of you, palms facing you, with your fingertips just slightly overlapping. This represents your sarcomere at rest. Now move your fingers inward, fingers sliding past each other. This represents a sarcomere contracting. As the filaments slide past each other and the sarcomere shortens, force is produced.

The Tortoise and the Hare: Fast and Slow Twitch Fibers

Some movement activities require endurance, whereas others require a lot of force over a short period of time. Fortunately, differences in motor units, both in terms of how the nerve functions and the chemistry and action of the muscle fibers, make both types of movement possible.

Muscle fibers are generally divided into two primary groups: slow twitch and fast twitch, although an intermediate category also exists. The *twitch* is a reference to the speed and frequency of the neural signal passing through the motor unit.

The fiber type you are born with may pick your best sport for you. Fast and slow twitch muscle fibers are not interchangeable, so what you are born with will, to an extent, explain a lot about the type of activity in which you excel. Born with a lot of slow twitch? Well, the 100-meter sprint may not be the event for you, but you'd probably make a great marathon runner! Lots of fast twitch? Chances are power sports are great for you. Of course, most of us have a combination of slow and fast twitch muscle fibers, enabling us to perform a wide range of sport activities.

Not too strong, but keeps on keeping on: The slow twitch muscle fiber

Think of slow twitch muscle fibers like a tortoise: They're not particularly fast, but they do keep on going. Slow twitch motor units have some common characteristics, outlined in Table 10-1.

Table 10-1	Common Characteristics of Slow Twitch Muscles
Nerve Characteristics	*Muscle Fiber Characteristics*
The nerve is small and reaches a threshold for firing with a small stimulus from the brain.	Slow twitch muscles have large numbers of mitochondria.
The frequency of the nerve twitch is slow, and the magnitude of the twitch is low.	They're aerobic fibers, capable of making ATP by using aerobic metabolism (see Chapter 4). They also use fats, carbohydrates, and lactic acid as a fuel source.
The nerve connects to only a few fibers (maybe 100–500 fibers per nerve), making it handy for fine motor control.	They have large amounts of *myoglobin*, an iron-containing protein that transports oxygen through the muscle tissue and gives the fibers a darker, reddish pigment. (Slow twitch fibers are sometimes called *red fibers*. If you're a turkey eater, the slow twitch muscles would be the dark meat!)
	Slow twitch muscle fibers are relatively smaller and weaker than their fast twitch counterparts.

Slow twitch muscles come in handy for any activity in which endurance is essential. Some muscles have a higher proportion of slow twitch fibers. For example, the soleus muscle, which is in the lower leg and is important for standing, has a high proportion of slow twitch fibers. In addition, because the nerve only connects to a few fibers, slow twitch motor units are usually associated with fine motor skills (like writing, typing, or blinking).

Big, strong, fast . . . and quickly tired: The fast twitch fiber

Fast twitch fibers, also called *fast twitch A* fibers, are like the hare: fast! These fibers are built for speed and for generating a lot of force. The downside is that they tend to fatigue quickly. Fast twitch fibers and their associated

---

Using slow twitch muscles to cool down!

Although slow twitch muscle fibers run on the byproducts of fat and carbohydrate metabolism, they have a unique enzyme that also allows them to use lactic acid as a fuel. The enzyme converts lactic acid back to pyruvic acid, which can then be taken up by the mitochondria. As the slow twitch fibers are activated, they use the lactic acid as a source of fuel, which helps clear away the lactic acid quickly. You can use this behavior to cool down.

Suppose that you're engaged in an intense activity, like running very hard for three to five minutes. During this activity, you build up a lot of lactic acid. What can you do immediately afterward to recover? Activate your slow twitch fibers by lightly jogging or walking. As you cool down with the light activity, you're clearing away the lactic acid much more quickly than you would if you just stand there with your hands on your knees!

---

nerves differ from slow twitch in a number of ways. Table 10-2 outlines their characteristics.

Table 10-2	Common Characteristics of Fast Twitch Muscles
Nerve Characteristics	**Muscle Fiber Characteristics**
The nerve is large and takes a higher level of stimulus to reach a threshold for firing. For light activities, the motor unit is generally not active.	Fast twitch fibers produce their energy by using anaerobic metabolism (Chapter 4). They don't use much oxygen.
The frequency of the nerve twitch is fast, and magnitude is large.	Anaerobic fibers produce ATP quickly, so fast twitch fibers can produce more force than slow twitch fibers can.
The nerve connects to many fibers (thousands of fibers per nerve). When the motor does reach a firing threshold, a lot of fibers are activated, and a lot of force is generated.	Anaerobic fibers produce more lactic acid, meaning they fatigue quickly.
	The fibers are large and respond more to strength training.

Fast twitch X, or intermediate, fibers

One slightly different version of the classic fast twitch fiber is the *intermediate fiber* (also called *fast twitch X* or *fast twitch B*). Although these fibers are still fast twitch, their chemistry is a little different. These fibers have some mitochondria, so they can get energy both from aerobic and anaerobic metabolism.

Fast twitch A versus X: Looking at training's effect on your muscles' behavior

Can jogging slow down your sprint speed? Maybe. Bodies are very sensitive to training and adapt to the type of training they receive. In the case of muscle fibers, the fast twitch X fibers may adapt to one type of training at the expense of adaptions to another.

Fast twitch X fibers can be trained so that they function more similarly to aerobic fibers or more similarly to anaerobic fibers. If you do a lot of endurance training, you'll improve your endurance, but you may lose some sprint speed. If you do a lot of sprint training, you could improve your sprint speed but lose some of your endurance. If you're a really serious athlete, consider blending your training to include a range of intensity so that all the muscle fibers can adapt and gain performance.

Working in Unison: How the Muscle Behaves

Muscles are much more than simple pieces of flesh that merely contract and relax. In this section, we investigate the properties that help define muscle function. How reactive a muscle is, how much it can stretch, and whether or not it returns to its normal length are all considerations that affect or dictate muscle function.

Looking at a muscle's response

Muscles produce the force needed for any type of movement, whether it be swatting a bug or running a marathon. Obviously, each action is unique and has particular demands. The force that a muscle or group of muscles can produce is dependent on its behavioral properties, outlined in the following list:

- **Extensibility:** *Extensibility* refers to the muscle's ability to be stretched or lengthened beyond its normal resting length. A good example of extensibility is how the quadriceps (the muscle in the front of the upper leg) stretch when the hamstrings are contracted to bend the knees.

- **Elasticity:** Whereas extensibility refers to the ability of a muscle to be stretched, *elasticity* refers to a muscle's ability to be stretched or elongated and then to return to its normal resting length afterward. When a muscle is stretched, it returns to its normal length unless an injury occurs from being stretched too far; in this case, the muscle is strained and the muscle fibers are torn.

✔ **Irritability:** In this context, *irritability* doesn't mean getting angry; instead, it refers to the muscle's ability to respond to stimuli. Also known as *excitability,* irritability describes the reactivity of a muscle and may dictate the timing and amount of stimulus needed for a contraction.

✔ **Contractility:** In all cases, the muscles must generate tension to create movements. The ability to create tension is referred to as the muscle's *contractility.* The amount of tension or how hard a muscle contracts depends on the muscle's length, its timing, which motor units get innervated, and the position of the joint.

Noting muscles' organizational structure

You may have heard that form follows function — a maxim that definitely holds true when it comes to your muscles. Muscles have a distinct organizational pattern and are connected to the bones in ways that enhance your mobility while simultaneously giving you strength. Read on to discover how muscle organization is key to movement.

The architecture of the muscle fiber

The architecture of your muscles — that is, their size and shape — differs from body part to body part and helps to dictate their function. Some muscles enable a large range of motion, and others have less extensive range of motion but provide support and stability.

Muscles are typically broken up into two structural categories: pennate and parallel (see Figure 10-4):

✔ **Parallel:** Muscles with parallel architecture have fibers that run in a parallel fashion along the length of the muscle. This fiber orientation is more conducive to large ranges of motion. The shapes that possess a parallel arrangement are fusiform, triangular, flat, and strap.

✔ **Pennate:** In pennate muscle architecture, the fibers are oriented at an angle into their tendons. Because of the larger number of muscle fibers that are attached to the tendon, pennate muscles are able to generate more force. In this category, muscles can be *unipennate* (attaching to one side of the tendon), *bipennate* (attaching to both sides of a tendon), or *multipennate* (attaching to the tendon in multiple locations), depending on the number of tendons and how the muscle fibers attach to them.

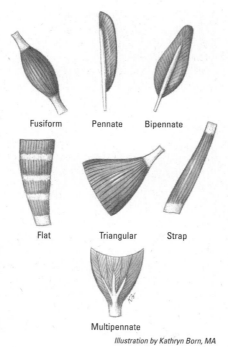

Fusiform   Pennate   Bipennate

Flat   Triangular   Strap

Figure 10-4:
Pennate
and parallel
muscles.

Multipennate

Illustration by Kathryn Born, MA

Origin, insertion, and lights . . . camera . . . action!

What the muscle attaches to is also a key component of the movement it creates when tension is developed. When you think about the muscle attachments, you need to know the difference between the origin and the insertion:

✔ The *origin* of a muscle is typically the less moveable portion of the attachment. Another way to think of the origin is that it is the more proximal attachment site of the muscle.

✔ The *insertion* of a muscle is typically the more moveable portion of the attachment and also the more distal attachment of the muscle.

✔ The *action* is simply the movement that is created in the joint when tension is created. What kind of action the muscle is capable of depends on the location of each of these attachments.

Two things dictate a muscle's function: where the muscles attach and at what angles they attach. Consider the knee, for example. A number of muscles attach to the knee cap (patella), and each attaches at a different angle. Some pull the patella up, and others pull it to the outside or inside. If all these muscles work together, the result is a coordinated upward movement of the patella when the knee is extended and a coordinated downward movement when the knee is flexed. The angle of pull, coupled with the amount of force from the muscles, dictates the muscles' effects on many of the joints. Figure 10-5 illustrates the different angles of pull.

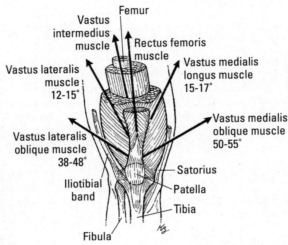

Figure 10-5:
The angles of pull.

Illustration by Kathryn Born, MA

Typically, muscles generate more than one action. Often the types of actions that a particular muscle contributes to depend on the body's position at the time. For example, your pectoralis major (chest muscle) contributes to both horizontal adduction (when the arm is parallel with the ground) and internal rotation (regardless of arm position).

Two-joint and multi-joint muscles

Most muscles span only a single joint and, as a result, provide a standard action with each contraction. However, some muscles span two or more joints, and their function is slightly more complex.

- ✔ **A two-joint muscle:** Take the biceps brachii (long head), for instance; this muscle originates off the scapula and inserts on the radius in the lower arm. Based on its attachment sites, it primarily supports either elbow flexion or shoulder flexion and is considered a *bi-articular* muscle (that is, a two-joint muscle).

- ✔ **A multi-joint muscle:** An example of a multi-joint muscle is the *erector spinae*, the muscle group that extends along your spinal column and spans multiple joints. When this muscle contracts, the action involves multiple segments of the vertebrae. During contraction, your back is pulled upright, and you stand up straight.

Because muscles often span more than just one joint, their function is largely dictated by the body's position when they contract. For instance, when you are bent all the way over, the erector spinae literally pulls you up (each vertebrae is involved), but when you are already up, it holds you there, keeping you upright.

Pulling harder and harder: Gradation of muscle force

As we explain earlier in the chapter, a motor unit is a single nerve and the few muscle fibers that it connects to. If only one nerve is stimulated and only the few fibers connected to it are contracted, you wouldn't be able to generate much force — a scenario akin to having only one person on a tug-of-war team: not strong enough to produce any sort of effective movement.

What is essential for most movement activities is a *gradation in strength,* that is, the ability to increase force output as needed, and the ability to get just enough muscle involved to do the job. Fortunately, the coordination between your brain and your muscles makes such gradation possible.

The key factors in generating muscular force are

- ✔ **The size of the brain (neural) signal:** As the desired force output rises, the brain sends a larger signal down the motor nerves. The size of the signal dictates how many motor units are activated. The larger the signal, the more motor units kick into gear, and the more power you're able to generate.

- ✔ **How strong the signal has to be for the motor units to fire:** Smaller motor units require less of a signal before they reach a threshold to fire. These smaller units are the first ones activated. Slow twitch motor units have the lowest threshold for recruitment, so they're used first. Activities such as walking, writing, standing, lifting a light weight, and so on probably use more slow twitch fibers!

A muscle fiber never "sort of" contracts. It either contracts, or it doesn't. Each motor unit has a threshold stimulus point. When the signal from the brain is less than the threshold, the motor unit doesn't contract. If the signal exceeds the threshold, then all the fibers associated with the motor unit contract. This all-or-nothing behavior of muscle contraction allows the size of the signal from the brain to control how many fibers are activated and gives your body excellent control of how much force you generate.

To generate more force, your body recruits more and more slow twitch fibers and then starts recruiting fast twitch fibers. As the neural signal grows, more slow twitch motor units are recruited. Just as if you were adding people to a tug-of-war team, the force generation increases. At some point, the neural signal becomes great enough that larger, fast twitch motor units are recruited as well. Usually, fast twitch A fibers are recruited first, followed by fast twitch X. Figure 10-6 shows the sequence of recruitment as the amount of required force increases.

Motor units are recruited in increasing amounts, based on the size of the neural signal. Slow twitch fibers are recruited first, followed by fast twitch.

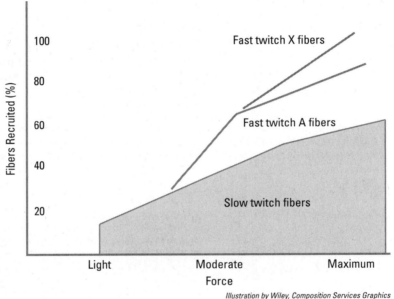

Figure 10-6:
More and more motor units are recruited as the required force goes up.

Illustration by Wiley, Composition Services Graphics

Acting on Instinct: Hardwired Muscle Reflexes

In any system, some decisions simply don't really need a lot of consideration (hence, the term "no brainer"). When it comes to movement, some sensory signals are directly linked to a motor response without first going to the brain. If you touch a hot stove, for example, do you really want your nervous system to take the time to send a signal to your brain and your brain to take the time to register the signal as pain and then take more time to send a signal to move your hand from the burning surface? Probably not!

Instead, some movements are reflexive; they don't require thought. Of course, not all reflexes are of the "yank your hand from a burning stove" variety. Some simply control how your muscles respond to different kinds of forces or stimuli. Reflexes are important to movement because they speed reactions and provide safety.

Developing tension

Tension and how it's created are at the core of muscle function. Although a contraction leads to movement, not all contractions are the same. How fast the muscle contracts, the types of contraction, and the mechanisms by which the muscle is contracted are integral to how you move. In this section, we explain two kinds of reflexes — the myotatic stretch reflex and the tension reflex — that impact how your muscles react to sudden or excessive forces.

Muscles spindles and the myotatic stretch reflex

The muscle has its own built-in reflex to help prevent the muscle from being overstretched. If, for example, someone or something tries to extend your arm really fast, you may notice that your biceps instinctively contracts — pulls back — to stop the movement. Without this reflexive action, the force extending your arm very quickly may injure you. This action and ones like it are dependent on the muscle spindle.

The *muscle spindle* is a *proprioceptor* (a sensory receptor) that resides within the muscle fibers and detects changes in muscle length and the rate of change. When the spindle is activated because the contraction is too fast, pushes too far, or has too much force, the myotatic stretch reflex is initiated.

The *myotatic stretch reflex* is a simple reflex that occurs at the spinal cord. When triggered, this reflex contracts the muscle being stretched, called the *agonist* (refer to Chapter 9). As the agonist contracts, the original force, velocity, and range of motion decrease, ultimately protecting the muscle from injury. If you attempt to kick a ball and miss, for example, the stretch reflex recognizes your blunder and stops you from going too far, preventing you from hurting yourself.

The Golgi tendon and the tension reflex

The last thing you want to happen when you're lifting something heavy or otherwise exerting yourself is to have your muscle pull so hard that the tendon actually pulls away from the bone it's attached to. Fortunately, your body has a way to stop the muscle from hurting itself: It's the Golgi tendon organ.

The *Golgi tendon organ (GTO)* is a proprioceptor that is located where the muscle attaches to the tendon. It responds to tension development, much as a spider senses movement on its web. The GTO has a threshold of tension that it considers "too much." When the tension exceeds that amount, the organ sends a signal to the spinal cord and then immediately to the muscle, inhibiting further contraction. Essentially, it turns down the muscle spindle to prevent or minimize additional contraction and keep a lid on any additional tension development. (You can read more about the GTO in Chapter 9.)

You can manipulate how the Golgi tendon works to attain your training goals:

✔ **Using it to stretch out your muscles:** When you perform a static stretch (refer to Chapter 9) and hold the stretch for a long period, you put tension on the tendon. What you may notice after about 30 seconds is that the muscle starts to relax. This is the Golgi tendon doing its job. It senses the tension you put on the muscle and then reflexively causes a reduction in muscle tension. So hold the stretch and watch the muscle relax!

✔ **Training it to increase strength:** One interesting aspect of the Golgi tendon is that you can actually change its tension threshold. With repeated bouts of exercise, the threshold moves to a higher amount, meaning that you can generate more tension (you become stronger!) using muscle you already have.

But watch out: Highly trained athletes have pushed the threshold so high that they put themselves at risk of tearing the tendon from the bone.

Shortening, lengthening, or not! Types of contractions

Most people equate muscle contraction to movement. Although often the case, this isn't always so. If you've ever held the door open for someone or simply tried to balance on one foot, you know that some activities require that your muscles fire even though your joints don't move.

Comparing dynamic and static contractions

Muscles act in both a static and dynamic way, providing support for and leading the way for movement. *Static muscle activity* results when a muscle or group of muscles contracts, but the joint doesn't move or moves only minimally. Typically, the purpose of a static contraction is to provide support for the joint or to minimize movement produced by an external force. This type of contraction is considered *isometric,* because the joint doesn't move.

In *dynamic contractions,* the joint and its related body part moves. Dynamic contractions make various types of movement possible, and they occur to control motion from various internal and external influences. *Isotonic contractions,* in which the joint moves, are the most common types of dynamic muscle activity. Isotonic contractions are broken up into either concentric or eccentric components, explained in the next section.

Considering concentric and eccentric contractions

When you think about a muscle contracting, you probably simply picture a muscle shortening and causing the movement. However, dynamic movement is more complex than that and is dependent on several muscles and/or muscle groups communicating with one another.

When one muscle shortens as it contracts, another muscle lengthens. Each — either lengthening or shortening — is a particular type of contraction:

- **A concentric contraction** occurs when a muscle shortens during activation.
- **An eccentric contraction** results when a muscle lengthens during activation. The job of an eccentrically firing muscle is often to resist the motion of another muscle or to control how the joint moves.

Movement involves a collection of muscles firing both concentrically and eccentrically. Think of the concentric contractions as the movement producers and the eccentric ones as the movement stoppers. Eccentric contractions are also more often employed to control the movement. A good example of controlling a movement is that of hand-eye coordination. If you reach out to grab a door knob and have only concentric muscle firing, you'll reach out toward the knob and strike the door. Nothing — that is, no eccentric contraction — fires to slow your movement down as you reach the knob and or enables you to perform this action in a controlled fashion.

Understanding the importance of concentric and eccentric contractions is easier if you remember that a muscle is never totally relaxed. Consider what goes on with your arm when you lift a dumbbell. Your biceps contracts and shortens as you flex your elbow. Simultaneously, your triceps contracts to control the amount and velocity of the flexion, but it does so as it lengthens.

Recognizing the different ways muscles work

When muscles contract, they typically do so by applying tension to the sites of attachment and, as a result, create movement. Depending on the factors like fiber architecture, behavioral properties, and types of contractions, all covered in earlier sections, the way the muscles work may also differ (see Figure 10-7):

✔ **Agonists and assistors:** The acting muscle responsible for creating a given movement is the agonist. Although any muscle that contributes to the movement is considered an agonist, some muscles have a greater impact on the movement than others. The muscles with the greatest impact are called simply the *agonists,* or *primary movers.* Those that contribute to the movement in smaller ways are referred to as *assistors.* For example, elbow flexion is accomplished when the biceps brachii (short and long heads), brachialis, and brachioradialis all contract; however, the biceps brachii is considered the agonist, or primary mover, and the others are considered assistors.

✔ **Antagonist:** The muscle that contributes directly opposite to the movement that's occurring is referred to as the *antagonist.* Typically, these muscles are located on the side of the joint opposite the agonists, and they cooperate with the agonists by relaxing during the movement.

The act of the antagonist relaxing or decreasing its gradation (refer to the earlier section "Pulling harder and harder: Gradation of muscle force") during the primary movement is usually an eccentric contraction. However, the key thing to remember is that the antagonist is the muscle that, when contracting concentrically (that is, shortening), provides the opposite motion. When it contracts eccentrically (decreases in gradation), it helps to control the motion and allows the movement that the agonist is trying to accomplish.

✔ **Stabilizer:** For a movement to occur, the muscles must act to stabilize, or *fixate,* the area so that the limb can exert the force needed to create movement. The muscles that act in this manner are the *stabilizers.* The abdominal muscles, or your *core,* are good examples. When you kick a soccer ball, for example, your core contracts and creates a firm point of support that enables the hip to move and strike the ball. If you don't have good abdominal support, you can strike the ball with only minimal force.

✔ **Neutralizer:** *Neutralizers* are the muscles responsible for controlling the influence that internal or external forces may have on the desired movement. Not all muscles that attach to a joint contribute in a way that facilitates the movement; some may actually impede the movement. Therefore, you need muscles around the joint that play a neutralizing role, counteracting muscles that fire in a way that would derail the intended action. Neutralizers also minimize the effect of external forces that can alter the intended movement and potentially cause injury.

Any muscle can play any of these roles, which are essential to allowing and supporting muscle function. With every movement, an agonist is necessary to create the movement, the antagonist is necessary to help control the motion and allow it to happen, and other muscles are necessary to control the internal and external forces and to create a solid base of support for the movement.

In the next two sections, we explain two factors that dictate how muscles work: the length of a muscle and speed of movement.

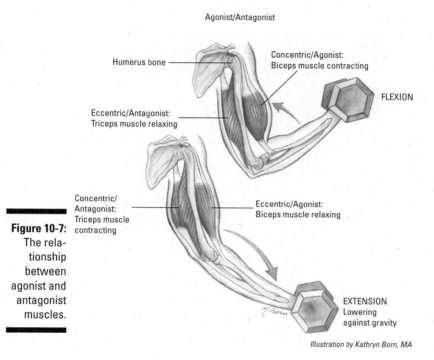

Agonist/Antagonist

Humerus bone

Concentric/Agonist:
Biceps muscle contracting

FLEXION

Eccentric/Antagonist:
Triceps muscle relaxing

Concentric/
Antagonist:
Triceps muscle
contracting

Eccentric/Agonist:
Biceps muscle relaxing

Figure 10-7:
The relationship
between
agonist and
antagonist
muscles.

EXTENSION
Lowering
against gravity

Illustration by Kathryn Born, MA

Examining the length-tension relationship

A muscle's length influences its force. The length-tension relationship explains how much force a muscle can create, given the muscle's length.

A muscle's optimal position for tension development is at approximately 120 percent to 130 percent of its resting length. Translation: If your muscle is stretched slightly beyond its normal resting position, it can produce more force. Conversely, when the length of the muscle is less than its normal resting length, how much force it can produce goes down because it has no more room to shorten.

Have you ever tried to jump up and touch a basketball hoop? If so, you undoubtedly squatted and then exploded up to touch it. This example shows how the length-tension relationship helps produce force — in this case, jumping higher. Heck, go ahead and stand up from reading this book and try to jump, first, without bending down first and then with bending. Which stance allowed you to go higher?

As you do activities on a regular basis, you begin to identify ways to make the tasks easier. Many of the changes you make relate to the length-tension relationship. You alter how much you twist or bend or reach for something, trying to make it easier. Pretty cool, isn't it?

Thinking about the force-velocity relationship

The speed, or velocity, at which a muscle contracts has a direct impact on the amount of force it can produce. More velocity produces less force; less velocity produces more force.

When the muscle is concentrically contracting (that is, shortening) and you attempt to move a relatively light object, you can do so easily and at a pretty high velocity. Think about throwing a baseball: The ball is light, and you can throw it really fast. Yet as the object you're trying to move becomes heavier, you are less able to move it as fast. Throwing a shot put, for example, is much more difficult than throwing a baseball, and you can't move it as fast because it's heavier.

When the resistance grows, movement becomes more and more difficult, as the concentric contraction turns isometric and then progresses to eccentric. Here's what happens:

1. **As the object continues to get heavier, you begin to transition from a concentric contraction, in which the muscle shortens, to an isometric one, at which point no movement is possible.**

   If you've ever been stuck in the middle of a dumbbell curl, you've experienced firsthand what an isometric contraction is!

2. **Finally, you transition to a eccentric contraction, the point at which the resistance is greater than the force being produced by the muscle, and the muscle begins to lengthen while it contracts.**

   An eccentric contraction has the ability to create a large amount of force, more than is typically seen with a concentric action. Also, while the resistance is significant enough to warrant an eccentric contraction, as the velocity increases, you start to see a significant increase in how much force the muscle produces.

Bottom line: Fast concentric contractions can't generate much muscular force, whereas fast eccentric contractions can create more muscular force.

Transitioning between forces: The electromechanical delay

During your normal daily activities, your muscles fire both concentrically and eccentrically, so it's only logical to imagine that a transition occurs between each. The *electromechanical delay* is that transition, and it represents the time it takes for a muscle to go from acting eccentrically to acting concentrically. This phenomenon is often referred to as the *stretch shortening cycle* or the *amortization time*.

Consider what happens when you squat down and then jump. As you squat, your quadriceps muscles lengthen (stretch), representing an eccentric contraction. As you go to jump, they fire concentrically (shorten) to boost you up. The period of time between the end of the eccentric contraction to the beginning of the concentric contraction is the electromechanical delay.

The delay needs to be a relatively quick one to ensure that the energy and elastic properties created during the eccentric portion of the activity aren't lost and actually end up enhancing the concentric portion. If you bend down and sit there for a few moments before you jump, you lose the benefits that the stretch or stored energy would have on the activity. Similarly, if the delay is really short, then the muscle isn't able to stretch to its maximal point of force production.

An additional concern with delays that are too long is that the joint is moved into a larger range of motion and injury may result — a situation that occurs in throwers. Research has shown that people who experience shoulder injuries actually have muscles with longer electromechanical delays. Their shoulders are forced to rotate farther than normal, often resulting in additional joint injury.

Plyometric training, a very common training technique in the sporting environment, is totally dependent on the stretch shortening cycle. This type of training often involves successive jumping or bounding, sometimes for distance and other times for height. By moving in this way, the athlete is able to maximize the force production within the muscles. The result? Higher jumpers and faster runners.

Training the Muscle to Work

Muscles are in a fairly constant state of change. If they are not being used at all (as happens, for example, when you're at bed rest due to the flu), they begin to *atrophy* (grow smaller and weaker). However, if you train the muscles, they can adapt, get stronger and faster, and even change the chemistry within them. "Use it or lose it" is a pretty accurate statement, and in this section, we explain how you can train the muscle to do more work.

Gaining the way you train: Specificity of training

Will being stronger make you faster? Although strength is certainly an important part of training for speed, your neuromuscular system is finicky. If you want to train for speed, you need to do speed training! As simple as this concept is, many people violate the *specificity of training* principle that says your training should be specific to the desired task. Some, for example, think — incorrectly, we might add — that swinging a heavy baseball bat gives them a faster bat swing or that being able to squat a heavy weight enables them perform a higher vertical jump. These kinds of violations of the specificity of training principle happen all too frequently. In this section, we explain the key principles of training specificity.

Training improvements are specific to the muscle fibers used

If you run a number of miles per week at a modest intensity, your slow twitch muscle fibers will be the primary fibers recruited for the activity (refer to the earlier section "The Tortoise and the Hare: Fast and Slow Twitch Motor Fibers"). So what happens during the big race when you need that big kick at the end to run at a higher intensity? Because you haven't trained for the additional fast twitch fiber recruitment needed for that kick, your kick will be tiny, and those who've trained more specifically for that moment will pass you by.

To avoid this scenario, incorporate additional training into your schedule that involves running at a higher intensity! Practice the kick to train your body to recruit the fast twitch fibers when needed. Then when the moment arrives, those fibers are ready for the action.

Improvements occur at the speed of training and all slower speeds

Slow speed strength training may help make a muscle strong, but it doesn't give it speed. If you want speed, you need to train the nervous system and muscles to function at speed. Improvements in speed training occur only at the fastest speed you train at. Training at faster speeds also improves your slower speeds, but you can never train slow and hope to be faster. So match at least some of your training to the speed of the activity!

Changing the load changes your speed

As we explain in the section "Thinking about the force-velocity relationship," force and velocity are related to each other. A light load can be moved quickly, whereas a heavy load is moved more slowly. If you want to swing a baseball bat fast, guess what happens if you pick up a heavier bat? Your swing is slower! Keep training with it, and you get better . . . at swinging a bat slowly. Your bat speed won't improve; in fact, you may actually slow down your speed. Yikes! The lesson? Use loads for training that are very similar to those in your sport.

Adaptations are specific to the joint angle and body position used in training

Functional movements need to be trained, and training needs to replicate as closely as possible the movements you want. If you alter the angles and body positions during training, you'll see those results when you perform the movements in competition. A good coach can help you learn the proper movements. Then it's up to you to train accurately!

I (Brian) used to teach martial arts. When I taught kids how to kick, I had the kids stretch their legs before kicking and, while doing so, imitate a perfect front kick. "You kick the way you stretch," I told them. "And no one likes a crooked kicker."

You need to train the chemistry of the muscle

As we state earlier, fast twitch and slow twitch muscle fibers differ in their chemistry. Slow twitch fibers are built for aerobic activity, and their enzymes are geared for aerobic energy production. Likewise, fast twitch fibers have anaerobic enzymes for ATP production. To train for more enzymes (and faster energy production), you need to train those fibers!

Fast twitch X (or intermediate) fibers can be trained for both anaerobic and aerobic energy production if you vary your training. Be sure to train for the type of adaptation you want to see in those fibers.

Making more muscle and gaining strength

Having more muscle mass has many advantages: Activities of daily living are easier, sport activities are possible, and injuries are less frequent. In addition, muscle burns calories, and stronger muscles lead to stronger bones. Muscle mass becomes even more important as you get older and start to think about things like osteoporosis and playing with the grandkids.

Fortunately, you can develop muscle through training. Muscle grows as a result of stress and damage — which sounds pretty traumatic, but don't worry: The stress is not that great, and surprisingly little effort is required to gain strength.

In fact, making muscle is a bit like making a callous. If you've ever played a guitar, you know that pressing the strings against the fret board is a bit painful at first. But with time, nice callouses develop on the fingertips, and playing becomes both easier and less painful.

Many different ways exist to give muscles enough work to grow and adapt. However, a few basic rules apply in all cases of strength training. Follow these rules (outlined in the next sections) to help in your training.

Rule 1: You must lift heavy enough

Weight training is not like aerobic training. Going for a walk is an excellent way to condition your heart and circulatory system. Your muscles, however, need more than that! Because stress and damage cause a muscle to grow (refer to the preceding section), you need to actually cause structural stress to your muscle. If you use loads that your muscles can lift with no stress, you won't see improved strength. The trick is to push your muscles beyond their comfort zone to cause enough stress without injuring the muscle.

How much is the right amount? As a general rule, you need to lift a load that is at least 60 percent of your *one repetition max (1RM),* the load you could lift just one time. For example, if you can lift 100 pounds one time only, your exercise weight should be about 60 pounds. Any lighter and the muscle easily lifts the load, and you don't achieve the necessary damage that makes adaptation and growth possible.

Eccentric contractions (when the muscle lengthens while contracting; refer to the earlier section "Considering concentric and eccentric contractions") seem to cause the most damage to the muscle during weight lifting. That's good! In fact, you should emphasize the eccentric contractions. So don't just lift the weight and then drop it: Lower it in a controlled manner to work the muscle during that eccentric phase.

Rule 2: You must lift to fatigue

Lifting a single load, even one that's heavy enough to damage the muscle, doesn't result in *all* your muscles being stressed. In fact, any load that can be lifted more than once is a *submaximal load* (your maximal load is one that is so heavy you can lift it only one time).

When lifting submaximal loads, on repetition 1, you may need only a fraction of your muscle. As those fibers become fatigued, your body calls up more fibers to help carry the load. This process — the used fibers getting tired and the body tapping other muscle fibers as reinforcement — continues with each repetition until all the muscle fibers are involved *and fatigued.* Only at this point, when all your muscle fibers are fatigued, have you worked the muscle, so don't stop early!

Rule 3: Growth happens during the recovery, so eat and rest

Muscles grow during the recovery time between workouts. The point? Recovery is very important to ensure muscle growth. Here are some things you can do to make the most of the recovery time:

- ✔ **Eat enough carbohydrates.** You need energy to fuel the growth. Carbohydrates should make up 50 percent to 60 percent of your diet to fuel the growth of new muscle tissue.

- ✔ **Include an adequate amount of protein in your diet.** Protein helps form the contractile filaments and connective tissue that make up the muscle. You should have between 0.4 and 0.6 grams of protein per pound (g/lb) of body weight. For example, at 0.5 g/lb, a 180-pound person needs 90 grams of protein per day.

- ✔ **Sleep!** During restful sleep, the hormones of muscle growth (growth hormone and testosterone) are highest. If you don't get adequate rest, these hormones won't be as elevated.

- ✔ **Base length of recovery time on the intensity of the training and take a break between workouts.** The harder the work, the more recovery you need. Light work, like gardening, can be done daily with little rest time needed. However, lifting a heavy load, like say 75 percent of your 1RM, causes stress and strain on the muscle and stimulates repair and growth. You can't train a given muscle daily because it needs the recovery time to actually adapt. The more overload you give muscle, the more recovery needed. Usually 24 to 48 hours is enough.

Rule 4: Progressively increase the load as the muscle adapts

Adaptation is great; however, after a muscle has adapted to a load, continuing to train the same way results in a lack of progress. For this reason, you want to engage in *progressive resistance training,* in which you progressively increase the loads as the muscle adapts. One way to track your progress is by the number of repetitions completed. If, for example, you choose a weight that is 60 percent of your 1RM, your muscles should reach fatigue after about 15 to 20 repetitions. If you have adapted enough so that you can lift the weight 23 times, it's time to add more load!

The amount of the increase varies and depends on the size of the muscle. Should you increase five pounds, ten pounds, or more? Rather than think in pounds, consider increasing by a percent of load. Doing so keeps the increase standardized across the different muscle groups. A 10 percent increase in load should be adequate to provide a nice progression in load.

Seeing how your body adapts to strength training

Strength training doesn't affect just the muscle fibers; it affects the brain and nervous system as well. Nervous system adaptations can happen very quickly, whereas growing muscle tissue takes a bit more time.

Nervous system adaptations

The nervous system adapts quickly to strength training and can result in a substantial increase in strength without any actual change in the size of the muscle fibers. The following neurological adaptations result in substantially increased strength within the first four to eight weeks of training:

- ✔ **Increased recruitment of available muscle fibers:** Untrained individuals can't access all the fibers they have for contraction. However, within just a few weeks of strength training, the body is able to recruit more muscle fibers for action. In other words, you can use more of what you always had!

- ✔ **Increased frequency of activation:** The brain sends signals for contraction to the muscle in pulses. Any increase in pulse frequency means that the muscles receive these signals more often. Because strength training increases the motor neuron firing rate, the muscle fibers spend more time contracting rather than relaxing.

- ✔ **Increased motor unit synchronization:** Just as in tug of war, you get the most force if everyone pulls together. Untrained motor units (refer to the earlier section "The motor unit: Connecting the nerve and the muscle") tend to have an unorganized firing rate: Some units fire while others don't. Strength training coordinates (or increases the synchronized action of) the motor units so that the fibers are all pulling at the same time.

- ✔ **Increased threshold for Golgi tendon activation:** The Golgi tendon, covered in the earlier section "The Golgi tendon and the tension reflex," is designed to inhibit additional muscle contraction when too much tension is detected. However, strength training changes this threshold for activation so that more force is required to invoke the reflex. In other words, the muscle can generate more force before it gets inhibited. Translation: You're stronger!

These adaptations benefit anyone who experiences them, but they greatly improve the quality of life for people who suffer debilitating weakness. Think of how a little added strength can help someone walk up stairs or get out of a chair!

Muscle tissue adaptations

Actual changes in muscle tissue due to strength training take a bit longer than neurological changes. However, by eight to twelve weeks, you should start seeing the following changes in muscle tissue if you follow the strength training rules outlined in the earlier section "Making more muscle and gaining strength":

- **Increased muscle size *(hypertrophy):*** The increase in size is seen in both the slow twitch and the fast twitch fibers. However, the fast twitch fibers seem much more responsive to strength training and grow the most.

- **Increase in contractile proteins:** Actin and myosin interaction is the foundation of muscle contraction (refer to the earlier section "The sarcomere and its parts: Shortening to produce force"). Having more actin and myosin means having more pulling power and a stronger muscle. Strength training increases the number of these proteins.

- **Increase in fast twitch X to A fiber transition:** Strength training is an anaerobic activity. As a result, when you engage in this type of training, the fast twitch X fibers become more like the high power fast twitch A fibers. Yes, this training may reduce your endurance capacity, but you gain strength!

Recognizing Sources of Muscle Fatigue

Sometimes the fastest person isn't the one who wins the race; it's the person who doesn't get tired as quickly. Fatigue is a limiting factor in many activities. You try to push yourself as hard as you can, and you try to hang on as long as you can before fatigue grabs hold of you. Depending on the activity, fatigue may come to the muscle in a variety of forms, which we explain in this section.

Running out of gas

As we explain in Chapter 4, the muscle gets its supply of ATP (energy) by metabolizing fuels. Running out of these fuels means fast fatigue:

- **Creatine phosphate:** This fuel is stored in the muscle for only about ten seconds of high-intensity work. When it runs out, the muscle fatigues!

- **Glycogen:** Glycogen is stored in the liver and muscle. You have about 2,000 calories' worth, which is enough for about a 20-mile run. When that fuel runs out, the muscle can't keep going!

Suffering from bad (lactic) acid

Lactic acid interferes with muscle contraction. Sarcomeres contract when calcium binds to the protein troponin, which helps open a binding site for the myosin to grab (refer to the earlier section "The Foundations for Muscle Movement: The Science behind Contraction"). Unfortunately, lactic acid (and the acidic hydrogen ion [H^+] that makes up part of it) competes with the calcium for the troponin binding site. As a result, the H^+ ion from the lactic acid prevents the calcium from binding to the troponin. Without calcium binding to troponin, muscle contraction is blocked. As acid levels increase during activity, more muscle contractions get blocked, and you start to fatigue.

You can recognize the fatigue that results from lactic acid as it happens. As you exercise and your muscle fibers begin to build up lactic acid, blocking contraction, you may notice that your running stride becomes "choppy," or your pedal rate starts to vary. What's happening is that the non-fatigued muscles are trying to help out with the movements, and your body uses different muscle activation strategies to do the same work. This situation can slow you down and also make your movements less precise.

More bad (lactic) acid: Slowing nerve conduction

Normally electrical signals move quickly along the axons of the nerve, skipping from node to node like a stone across water (Chapter 3 has more on the nervous system). However, when H^+ ions from lactic acid build up, the signals slow down. If you're trying to coordinate a movement, as you do when you run, jump, or shoot a ball, for example, you may notice that the sequence of your movements starts getting thrown off a bit. The reason is that the signals just don't quite get to the muscle in time, resulting in fatigue and reduced performance.

Getting the message from your brain to stop

One last cause of fatigue can be your own brain. Lactic acid and the resulting acidity (H^+ ions) can cause you to feel tired and in pain — and even nauseous — and can cause you simply to stop. Because motor activity is linked to sensory information (refer to Chapter 3), physiologic pain can result in a reduced motor output to the muscles. A reduced output from the central nervous system results in less effort (like less load lifted or slower running speed) and, therefore, fatigue.

Athletes have been trained to push well beyond the normal limits at which most of the rest of us would fatigue, even to the point of injury. To some degree, this is a good thing, but it can be taken to extremes. Pushing beyond what you used to think was fatigue may get you to perform better (like finishing that last mile of the race). Pushing to the point of weakness and leading to injury (like runner collapsing from exhaustion during a marathon) can mean you won't finish the race. The trick is to understand the limits of your body's performance.

Being a Pain: Muscle Soreness and Injury

People often train and compete at the limits of their physical abilities. That level of effort sometimes results in trauma to the body. You may notice some of the outcomes right away; they may even cause you to stop the activity. Other outcomes may take a day or two to hit, and again, the pain can limit your performance. In this section, we explore the possible sources of both *acute pain,* which happens during the activity, and *delayed onset pain,* which seems to happen later.

It hurts right now! Acute muscle soreness

Pain during exercise can come about when pain receptors are activated in the tissue. Your body is sensitive to changes in levels of acidity as well as pressure changes in the tissue. As a result, when you work hard, you'll experience some pain. The good news is that this pain goes away fairly quickly after you stop. The next sections describe the two big culprits of acute muscle soreness.

Lactic acid . . . again

As high-intensity exercise continues and the anaerobic system is forced to generate ATP, lactic acid begins to accumulate. It may also accumulate because blood flow to a tissue can't keep up with the demand for oxygen. In addition to causing fatigue, lactic acid also causes pain, and it may discourage you from continuing the exercise. Some athletes can push past this pain and keep going (although they will still fatigue).

Immediately following the exercise, your body starts to get rid of the lactic acid as your liver converts it back to glucose. If you keep moving but at a slower pace, your slow twitch muscle fibers use the lactic acid as a fuel, which is another way to get rid of it. In fact, you can even breathe off the lactic acid, because some of it is neutralized in your blood and converted to CO_2. So keep moving and keep breathing! Recovery only takes a few minutes.

Edema

Imagine your biceps muscle about to contract and lift a heavy weight. Now imagine a blood vessel moving through the muscle. What happens to the blood vessel when the biceps contracts? Pressure! This pressure can actually force the *plasma* (the watery portion of blood) out of the blood vessel and into the space between the muscle cells (called the *interstitial space*). While this phenomenon may give you a "pumped up" feeling, the interstitial pressure may also cause pain!

Your lymphatic system eventually drains the fluid in the interstitial space back into your blood, but that process can take between 10 and 20 minutes. At least you look good and pumped up as you wait for the pain to subside!

Delayed onset muscle soreness

Have you ever exercised, only to find that two days later you're sore all over? Why does the pain not occur until two days later rather than right away? Delayed onset muscle soreness (DOMS) is still a bit of a mystery, but a few theories exist, based on evidence available so far:

- **Muscle spasm theory:** Muscle is excitable tissue. When the brain sends stimuli, the muscles contract. Some researchers speculate that, if you work a muscle hard and that muscle doesn't get much oxygen and is surrounded by lactic acid, it becomes irritated and more excitable — a situation that causes spontaneous contractions, or spasms. Muscle spasms can be detected both by feeling the muscle (it feels hard and contracted) and by measuring the muscle's electrical activity (electromyography shows that it is contracting).

    Treatment for this type of soreness is to perform slow, static stretches. Hold the stretch of the muscle for 30 to 60 seconds. Doing so induces the Golgi tendon to reflexively inhibit the contractions and relax the muscle. Refer to the earlier section "The Golgi tendon and the tension reflex" for details.

- **Connective tissue damage:** If you stand along a river and see parts of a house float by, you probably have a pretty good sense that something bad happened upstream. Same in the body. When parts of proteins that belong in the muscle are found floating in the blood, then some type of damage has occurred.

    Collagen is a connective tissue within the muscle that helps hold the muscle together. When the muscle is damaged, the collagen is broken down, and the liver converts it into something called *hydroxyproline*. The presence of hydroxyproline, revealed by a blood test, seems to peak about 24 to 48 hours after heavy exercise.

✔ **Skeletal muscle damage:** If the damage to the muscle is great enough to destroy the sarcomere, then all sorts of things get dumped into the blood, including enzymes for energy production that are usually in the muscle (creatine kinase) and proteins. The presence of these things signals major damage to the muscle, and the levels peak in your blood 24 to 48 hours after heavy exercise.

Researchers hypothesize that the actions of repair *(macrophages)* stimulate nerve endings and cause pain and inflammation. Following the initial heavy exercise, soreness seems to go away, which is quite interesting. Why don't you get sore after every bout of exercise? Research is still looking for the answers.

In the meantime, few treatments are available to alleviate the pain from DOMS other than healing time. Menthol-based analgesic creams do seem to reduce some of the pain, so you might try some to take the edge off the pain.

When soreness is bad on the kidneys: Rhabdomyolysis

The kidneys are essential for removing waste products that accumulate in the body. Often your liver helps out by breaking products down to sizes that are easy to pass through the kidneys. However, in severe cases of muscle damage, the very large protein *myoglobin,* the iron-containing protein in the muscle, is dumped into the blood.

The kidneys have great difficulty passing this protein, so much so, in fact, that the kidney can shut down because of it, a condition called *rhabdomyolysis.* Rhabdomyolysis has led to death after heavy bouts of exercise, usually when the conditions were hot and the exerciser was dehydrated. (A small percentage of those who take the cholesterol medication called a *statin* also have an increased risk of muscle damage and rhabdomyolysis, and the risk increases with higher doses.)

If you suspect rhabdomyolysis (the urine will be noticeably dark due to the myoglobin), don't mess around! Get to the hospital, because kidney failure can happen quickly. Treatment is simple: Water, and lots of it!

Chapter 11

Motion Analysis: Seeing How You Move

Analyzing human movement is something we bet you do all the time, although you probably don't think of it that way. If you've ever "people watched" while sitting at the mall and said to yourself, "Poor girl, she's really limping," or "Wow, that jogger is certainly bouncing down the street," you've been analyzing movement.

As it turns out, an entire discipline is devoted to just this task. In this chapter, we explore a movement analysis process that makes evaluating movement easier. We also share some of the important principles and concepts to remember when you do so.

Investigating Movement: The Basics

Some of the most common reasons why people use motion analysis is to improve performance (they want to get bigger, stronger, or faster, for example) and to prevent or treat injury. To adequately analyze movement to achieve these goals, you need an approach that follows a defined process. This process, combined with information about the actual task, enables you to discover some very helpful and valuable information.

One of the biggest pitfalls related to motion analysis lies with the practitioner who, after eyeballing a task in real time, makes judgments without having knowledge of the specific task or its purpose, and without really understanding the person being studied. Each person is a unique individual, with his or her own past histories of injury, levels of strength, and previous training.

Choosing an approach

Motion analysis typically takes one of two forms: either quantitative or qualitative. Occasionally, a mixture of the two is used.

Each type of analysis requires the examiner to have a sufficient level of understanding of the task at hand and the intended outcomes of the performance. And each is beneficial to motion analysis efforts and possesses its own benefits and drawbacks, which we explain in the next sections.

Qualitative analysis

A *qualitative analysis* offers general motion analysis assessment and feedback pertaining to the movement. A pitching coach, for example, may notice that one of his athletes is really leaning to the side during the pitch. The coach may ask the athlete to straighten up and direct his body toward home plate when throwing and to avoid leaning so much. This type of feedback is considered qualitative.

Qualitative assessment is the most common type of assessment for these reasons:

- ✔ **It doesn't require high-tech instruments.** A qualitative analysis can be made simply by watching the movement, either in real time or on some type of video device for replay.

- ✔ **It doesn't require a particularly high level of skill.** As long as the examiner knows how the body can move, knows how it should move optimally to perform the task, and can pick up the key positions, he or she can complete the analysis.

In most cases — when a quick look is needed — a qualitative assessment is appropriate. Typically, this is going to be the only way a Little League coaches and physical education teachers may be able to examine their athletes or students. Whether watching the movement in real time or watching a replay, people can carry out qualitative assessments pretty easily and quickly.

Quantitative analysis

Quantitative analysis involves using actual measurement to gauge performance. If, instead of providing general assessment and guidance, the pitching coach records the motion with a video camera, uses analysis software in his evaluation of the movement, and then says to the pitcher something like, "You're leaning 18 degrees when you release the ball. You need to straighten up so that you don't lean more than 5 degrees," the coach is using quantitative analysis. In this case, the specific measurements that were made of the motion inform the coach's recommendations.

When the clinician needs more information about the amount of force, the velocity, or the angles of a movement — information that a qualitative analysis just can't provide — then a quantitative analysis is the better option.

Knowing types of tasks and feedback mechanisms

Motion analysis can be overwhelming if you don't really know how to go about it and don't know what to look at. If you're a novice, you may find yourself staring awkwardly at your subject, not really knowing where to begin. Unless you have an established process that includes knowledge of the person completing the task, the components of the movement being evaluated, and an understanding of feedback mechanisms, your analysis isn't likely to provide much useful information.

In the following sections, we explain what you need to know about the person you're evaluating and how understanding the different types of tasks (simple and complex) and feedback mechanisms (open and closed) is vital to making a sound evaluation and providing useful feedback. Head to the later section "Breaking Down an Analysis Model" for a process you can follow when the time comes to put your knowledge to work.

Comparing simple and complex tasks

Not all movements have the same difficulty or require similar amounts of preparation to execute. Simple movements require very little or no thought, and they aren't dependent on another body part completing its movement first. A good example of a simple task is picking up the phone.

Complex tasks, on the other hand, require thought and regular practice. They're made up of several individual components that must be completed in a particular way to produce a successful end result. An example of a complex task is pitching; it involves a coordinated effort between the legs, torso, and upper extremities. Each activity required for a successful pitch depends on another part of the body doing its job. Consider this sequence of movements:

1. **The pitcher pushes off his back foot to move his body forward.**

2. **His hips begin to twist to facilitate the slinglike motion at the root of throwing.**

3. **While his body twists from the hips up through the torso (core), he moves the ball forward and hurls it at the target.**

If any flaws occur within the chain of activity, the pitcher's performance and potentially his health may be in jeopardy.

Tommy John

Surgery to repair an injury to the *ulnar collateral ligament (UCL),* the ligament in the elbow, has been referred to as *tommy john* surgery since Dr. Frank Jobe first performed it on Los Angeles Dodgers pitcher Tommy John in 1974.

In this surgery, which was the first of its kind to attempt to combat what had previously been considered a career-ending injury, Dr. Jobe weaved a muscle tendon through the elbow joint to mimic the UCL. The idea was to stabilize the joint enough to allow Tommy to throw again. It worked. Tommy returned to the Dodgers and resumed pitching at a fairly high level. Since that first surgery, tommy john surgery and procedures like it have saved the careers of oodles of professional and amateur baseball pitchers.

Although many factors contribute to this injury, none is more harmful than poor biomechanics. When you throw with your limb in less-than-ideal positions, the soft tissues that typically support the joint break down and eventually fail or tear. Unfortunately, this injury that was once seen only in professional athletes now plagues many young players. The change is attributed to these main reasons:

✔ Many coaches don't understand what good throwing mechanics are. And if they don't know, they aren't able to intervene when a young athlete needs it.

✔ Kids who play on multiple teams and in multiple leagues are asked to throw all year round and never get a break. This constant use gives them no time to recover and heal.

So for all of you out there who think your 10-year-old is going to be the next Major League Cy Young award winner, how about giving the kid a break so that he actually has a chance to be?

Paying attention to open- versus closed-feedback mechanisms

Some tasks are difficult to alter while they are being performed; others are easily changed. The type of information that the body has to process and the way in which it processes that information (its *feedback mechanism*) determines whether an activity can be changed midcourse. There are two types of feedback mechanisms:

✔ **Open-feedback mechanism:** A task that has an open-feedback mechanism (or *open task*) can be changed in real time. During this type of movement, the body collects information and then alters the movement to maximize its success in that particular situation. For example, if you're going up for a jump shot and someone makes contact with you, you are able to change how you shoot the ball in midair.

✔ **Closed-feedback mechanism:** A closed-feedback mechanism task (or *closed task*) can't be altered after it's started. Closed tasks are those that, once turned on, can't be turned off. A free throw is a good example. The only way to be more successful with free throw shooting is to practice it over and over and over again.

Breaking Down an Analysis Model

To make analyzing movement easier and more efficient, you use a defined analysis model. A typical model for analysis involves knowing the nature and objective of the movement, observing and evaluating the movement, and providing recommendations based on the performance outcomes. We outline the different parts of the analysis mode in this section.

The outcomes vary extensively and can include information to help achieve different goals, like improving performance or avoiding injury. They can also focus on the whole task (how to better jump for maximum distance, for example) or specific components of a given task (informing a pitcher, for example, that he may be experiencing elbow pain during a pitch because he drops his arm slot when throwing).

Gaining background knowledge

As a basis for the beginning of the movement analysis, the examiner (coach, clinician, or personal trainer) must have some background knowledge about the task to be completed. Understanding what the performer is trying to accomplish and having an understanding of the components needed to be successful are essential to the analysis. Background knowledge, explained in the following sections, helps you to identify the key elements of the movement that need focus.

How to perform the task

You need to know the mechanics of the movement, its purpose, and the most efficient and safest ways to go about it. This knowledge makes you a more effective assessor.

Although having performed the task yourself can sometimes be helpful, it's not essential. Some of the best coaches weren't the greatest or most successful athletes; instead, their study of the game and its elements enables them to be successful. If you haven't personally performed the task you're evaluating, make sure you perform sufficient research into the mechanics and purpose of the task.

Many folks that analyze motion or provide coaching do so without considering how and what information they are basing their advice on. A clue that you're dealing with one of these people is someone who says, "I know good mechanics when I see them" and then, when asked to define good mechanics, isn't able to do so.

The objective of the movement and each of its components

The objective of a movement may be singular and relatively simple — performing a dumbbell curl, for example, which simply requires bending the elbow and repeating. Or the movement can *seem* simple, like getting the batter out, but is actually quite complex. Consider the individual components of a pitch:

- ✔ **How the pitcher stands on the rubber as he prepares to throw:** The stance provides the foundation for what's to come. Good balance and a controlled windup allow for a successful throwing motion.

- ✔ **How the pitcher moves his torso, hips, legs, and arms as he leads up to releasing the ball:** These motions generate the force leading up to his letting go of the ball.

- ✔ **How the pitcher moves his torso, hips, legs, and arms during the pitch itself:** These movements help dictate where the ball is going and what kind of movement (curve, slider, and so on) it may have.

- ✔ **How the pitcher slows everything down after the pitch:** At this point, the muscles have to stop the arm. If the pitcher has his body in the right position, some of the pressure is taken off of the shoulder muscles.

To accurately assess a task and ultimately provide feedback for improvement, you must understand the end goal and the individual components that make up that task.

The specific attributes of the performer

You must consider the level of skill, strength, range of motion, age, gender, and fitness level of the person you're evaluating. Each of these characteristics not only has a significant impact on how a particular task is executed, but it also determines whether the athlete is actually even able to perform the desired activity. For example, someone who lacks good balance isn't ready to be thrown up on the balance beam to do a routine. Instead, the athlete needs to be coached on specific techniques aimed at improving balance, like having him practice balancing in various positions. Such tasks enhance body awareness and allow the athlete to be more efficient and to avoid crazy movements that distract from the intended purpose.

Unfortunately, folks are too often put in situations that only lead to failure or injury. When I (Brian) see patients for elbow or shoulder pain, I invariably find out that they don't have the core or back strength needed to safely complete the task — like throwing a shot put — repetitively. By taking a systematic approach to preparing a person for a particular activity — helping the athlete develop the necessary strength, dexterity, or agility, for example — you can often avoid a bad situation.

Be sure to ask what the athlete's individual goals are. To improve perfor-mance? Avoid injury? Reduce existing pain or rehabilitate after an injury? The evaluation you perform for an athlete with performance-related goals is very different from the evaluation you perform for one who wants to reduce her pain. For the athlete recovering from an injury, you not only examine the actual movement but also the training regimen and number of rest days, whereas for the athlete just wanting to get more movement on her rise ball, you focus on the mechanics of the movement to determine whether an adjustment needs to be made.

Observing the subject in action

Obviously, you can't make an assessment of a movement, let alone provide feedback, without seeing the person perform the task. In this section, we tell you how to make this observation and what to look for during each phase of the movement. Depending on what you're looking for, you may use a power-and-return model, a three-phase model, or a model that breaks the move-ments down even more. The following sections go into more detail on the power-and-return and three-phase models, ones that you'll use most often.

Three-phase model

To make sense of what you see, you need to break the movement up into vari-ous segments, or phases. All complex movements require *preparation, execu-tion,* and *follow-through* components, and each of these phases has a series of movements that must occur in order for the next phase to follow and/or be successful. When you break a movement into these components, you're using the *three-phase model:*

✔ **Preparation phase:** Before each movement or task, preparatory work needs to be done. (For the purposes of this discussion, the preparation that we're talking about is the actual activity, not previous training and that sort of thing.) Getting into a position that facilitates the impending movement is the key to this phase. Proper preparation provides a firm foundation on which the athlete can perform the task competently and safely. To prepare for a standing long jump, for example, you squat down by bending at the knees, hips, and ankles — a position that maximizes the muscular force output (refer to Chapter 10 for information about the length-tension relationship). This stance helps you push off in a way that lets you jump as far as you can.

Other important considerations include timing and positioning on the court, field, or track. For example, the angle at which you release the javelin is also critical to successfully executing the task. Although it may not be directly related to the task (jumping) itself, it's closely related to the success or failure of the event.

- ✔ **Execution phase:** During this phase, the athlete's body executes the movement. He propels the object or jumps up and completes that triple axle on the ice. The appropriate timing, strength, and flexibility are required to maximize his performance and avoid injury.

- ✔ **Follow-through phase:** During this phase, the athlete slows down and returns to a normal (starting) position. The follow-through often requires deceleration of the limbs and continuation of the momentum as the body transitions to another task. For example, when you kick a soccer ball, your leg doesn't just stop after you strike the ball. Instead, it continues to move forward as your body follows and transitions to a step, typically followed by running.

Power-and-return model

Some professionals use a *power-and-return model* (also called the *two-phase model*). This model is broken down into . . . you guessed it . . . the power and the return phases. The *power phase* includes the execution of the task, and the *return phase* encompasses the preparatory phase. (***Note:*** Despite the name implying that the power phase happens before the return phase, remember that it does not.)

This model typically doesn't offer insight into the follow-through of the task as the three-phase model does. If your purpose is to evaluate the outcomes of the task rather than the intricacies that make it up, the power-and-return model may be sufficient; however, if your goal is to look deeply into the task, a three-phase approach is more appropriate.

A step-by-step guide to observing movement

Follow this process when observing the movement (remember, what you look for depends on why you are doing the evaluation or what the objective of the performance is):

1. **Break the activity up into phases with clear beginnings and endings, if possible.**

   Remember, most activities are complex in nature; they consist of many movements that need to be sequenced in a particular way. Identify the different phases and break the activity into either two, three, or, if needed, more phases.

   Looking deeper into the movement, most examiners also break up each individual phase and define what should happen within each. For instance, the gait cycle of walking consists broadly of the swing and stance phases. However, many biomechanists may break those phases up even further to include pre-swing, heel strike, weight acceptance, push off, and so on. In this case, more detailed assessment occurs, but this strategy isn't typical for most folks or tasks.

2. **Examine the participant as he performs the movement and note whether he moves through all the phases and whether he does so in the correct manner.**

   For example, if you're evaluating someone throwing, look to make sure the elbow and shoulder are doing what they're supposed to do when they're supposed to do it. Comparing what you see with the established norms (which you know either by personal experience or through research), you identify flaws, which can help you determine whether more complex issues exist. A complex issue commonly seen in throwers, for example, is that they don't move their hips early enough, which usually causes the arm to drag behind as the body moves forward. The end result is a lot of stress on the elbow and less velocity.

Exploring every aspect of each and every task can be daunting and so time-consuming and complex that it's pretty much impossible. To avoid this trap, try these strategies:

✔ **Answer the particular question at hand.** Taking a targeted approach ("Why does my ball hook every time I hit the golf ball," for example) can provide the necessary focus.

✔ **Break the motion up into the appropriate phases, based on your purpose.** By incorporating the key elements within each phase, you can appropriately and efficiently analyze their sequencing.

✔ **Identify common flaws.** Experts typically have their own abbreviated lists of movement flaws (stance, shoulder positioning, wrist rotation, and club alignment, for example) that they look for. Usually these common flaws or critical elements have been proven to cause other issues, like poor performance or injury.

A pitfall that many novice examiners fall into is that they observe the task from only one vantage point, even though most tasks occur in multiple planes of motion (refer to Chapter 7). For example, if you sit behind the backstop, you can see the front of the pitcher and he may look just fine, but when you examine the same pitcher from first base, you may notice he seems to have an extraordinarily long stride. Without witnessing the motion in all planes, you cannot make a full assessment. To avoid this trap, look at the task from each of these vantage points. Identifying flaws in any of the three planes of movement is the only way to fully evaluate the movement.

Making your evaluation and diagnosis

Ultimately the purpose of the motion analysis is to correct or improve the performance or avoid injury. To do so, you evaluate the subject's performance, identifying specific flaws and making diagnoses. You may notice, for example, that while shooting a free throw, the athlete doesn't appear to be bending his knees as he should. By evaluating critical elements of the free throw, you're able to diagnosis the movement flaw as a lack of knee movement.

The evaluation process typically involves comparing the athlete's actual performance against pre-defined critical factors. If the performer repeatedly falls outside the normal range for that factor, you note that finding, and intervention can then follow. Read on for things to keep in mind as you perform your evaluation and make your diagnosis.

Being mindful of personal differences

One of the difficulties that exists with motion analysis in living things is that each person is unique. The organic influences of the actual performance — height, strength, previous history of injury, and so on — differ from person to person. Because of these variations, no one looks exactly like anyone else when completing the same task. Even the critical elements of a particular task can vary a bit, which is why a range exists in normal subjects.

Just because someone is successful or pain free doesn't mean that person's technique is the right one to emulate. For example, you may think that you should model pitching mechanics after Major Leaguers. After all, they're successful and have lots of media attention. Yet despite being incredibly effective pitchers, many Major League pitchers are fighting mechanical flaws that ultimately have led to career-ending injuries. If you emulate them, you can find yourself promoting a technique that leads to repetitive injury that could ultimately be the end of a pitching career! Remember, mechanics are highly specialized in that individual and don't always translate well.

Tying recommendations to the purpose of the analysis

Preventing injury, enhancing performance, and overcoming discomfort are some of the reasons why people have their movements analyzed. Each of these reasons requires a unique look into the task, and the diagnosis and intervention recommendations should be related to the analysis's purpose. "How do I jump higher?" is a very different question from "Why does my knee hurt when I jump?" and requires the examination of a different set of key elements (even though, at times, they may be related).

Taking into account repetition and situation variability

Take a look at multiple repetitions in both practice and game situations. Examining someone during a practice yields different information than you get if you examine that person during a game. Also, if you take a look at only one repetition of the task, you can't be sure that what you see is actually what the athlete typically does. What you see between the scenarios is sure to surprise you — and make your analysis more accurate and helpful.

One thing you are sure to experience is that of *analyst bias.* We all have our own ways of looking at things and our own perspectives that impact what we think constitute critical factors and optimal performance. Analysts who don't take into account the whole person during an evaluation and who try to fit everyone into the same model are missing the boat. Cookie-cutter analysts just don't get the big picture of the individual.

A true expert identifies the unique factors of each participant and paints a full picture of both that person's performance and his or her flaws. In the end, the well-researched practitioner who stays up-to-date on the most current literature and trends in performance analysis is best able to provide the most robust assessment.

Providing intervention and feedback

What good is coming up with a diagnosis of the flaws in someone's mechanics if you don't provide feedback and intervention strategies? Saying what's wrong with someone's motion is generally easy; figuring out how to correct it is a bit more challenging. Key to providing feedback and intervention is having deep knowledge of the task at hand; relevant information about the participant's strength, injury status, and performance; and an understanding of the participant's goals.

Giving feedback

Based on what you know about the client, you can prioritize the feedback you give. Basically, you want to make deliberate decisions about what information to share and when. Here are some things to think about:

- **Prioritize the information.** Your client may not be able to absorb all the information or correct all his flaws at one time; therefore, you need to decide what information to share and when. For example, when I (Brian) do mechanical analyses on patients with multiple critical flaws, I provide feedback only on the most critical flaws and/or the ones that, if remediated, can help improve other aspects.

- **Organize the feedback in specific phases or in an order that facilitates growth or improvement.** Some tasks require significant practice and a step-by-step approach to be accomplished. As you decide what information to provide your clients, consider things like the best way to improve a skill, the level of difficulty of the different components, which tasks support others, the sequence in which the tasks are performed, which actions are key to preventing injury, and so on.

  For instance, say that you want to attempt a new dive off of the high jump; you don't just climb up the ladder and jump off. Instead, you spend lots of time on the ground, working on the phases and the motion sequences to build up the strength or range of motion necessary to complete the dive. After the ground-level work, you progress to a trampoline, and then you go into the pool. Considering a progression of skills is important when you establish plans for integration back to the activity.

✔ **Provide examples, where necessary.** For improvement to be successful, the participant must have proficiency in the movement, know what's expected, and know how the movement is broken down. Providing solid examples like video or still images is invaluable.

With tablet applications and digital image sharing, providing these examples is incredibly easy. Make sure, however, that your clients truly understand what you're sharing. If you use visual models, consider walking clients through what you're looking at on-screen to make sure they're clear about what you're pointing out.

✔ **Speak the lingo.** Every sport has its nuances and technical jargon, and part of your effectiveness as an examiner depends on your ability to speak the lingo. Doing so helps you build trust and give the participant confidence in your recommendations.

Intervening to get and keep them on track

Several things can affect athletes' ability to improve their performance. In this section, we look at the most important ones.

Encouraging the client's commitment to improving

Clients must want to improve and be willing to put in the necessary time and effort. What may seem like a simple change in the positioning of the arm when throwing can actually require tens of thousands of repetitions to correct permanently. (And just when you think your patients have the problem fixed, they revert back to old habits when they have to rush a throw or get tired or lazy!)

Engage patients and help them see the importance of improving their performance. Being told that they just aren't as good as they may have thought or that they're at risk of significant injury can be overwhelming or crushing emotionally. Some will just quit and move on; others will be so overwhelmed that they won't know where to start. That's where your encouragement and feedback is vital.

Facilitating opportunities for practice

Being able to make adjustments to someone's workout regimen or team practice may be invaluable to correcting organic issues early. For example, you may be able to add a practice drill that everyone does, or you may be able to purchase training equipment that influences a particular feature.

When helping a patient change the way he performs a given movement, the repetitions need to be coached appropriately, and flaws need to be identified immediately. You can't teach someone how to walk again, for example, without watching the client in action and providing real-time feedback. Because you may not be able to be present for every practice or game, the most effective way to achieve this objective is to work closely with the parents or coaches who tend to be present for every practice or game.

Studying Motion Analysis Examples

In this section, we include a few common tasks to demonstrate how movement analysis works. By going through both simple and complex tasks, you can see how to apply motion analysis to activities you experience regularly. The model we use here breaks down the activities into phases, as well as identifies the joint actions within each. For a review of the different analysis models and the phases in them, refer to the earlier section "Observing the subject in action."

Note: Some models go as far as listing the muscles that cause each joint action and explaining how those muscles work. Although this information can be very valuable to the examiner, especially when giving feedback, for our purposes, we just stick to the phase breakdown and the individual movements that cause each.

By using the processes presented earlier in this chapter and following the examples we include here, you should be able to complete your very own motion analysis. Of course you'll need to decide what the goal is (performance enhancement, for example, or injury prevention). But after you decide on those factors, you'll be on your way to being a full-fledged biomechanist. Here are some activities you can give a try:

- ✔ Making a free throw
- ✔ Standing from a seated position
- ✔ Shoveling dirt
- ✔ Swinging a golf club
- ✔ Making a tennis serve

Analyzing a squat

The squat is considered a key exercise in the sporting world because it strengthens a lot of the lower extremity muscles. A number of sporting activities are jumping and running related, so this exercise is a good one. However, if done incorrectly, injury can result.

In Figure 11-1, you can see how a relatively simple task can easily be broken down, using the power-and-return model. In this case, the analysis begins when the athlete is in the upright position and gets the bar off the rack. (The act of lowering to a full squat position can also be referred to as a *preparatory activity.*) In the power phase of the squat, the athlete rises up from a squat position, and in the return phase, he lowers back down to a squat. Here are the details:

✔ **In the power phase:** The individual joint motions in this phase include knee extension, hip extension, back extension, and ankle plantar flexion. Because the back doesn't really move during this phase, its action is isometric; the other actions are all moving.

✔ **In the return phase:** This phase is generally the opposite to the power phase. The knee flexion, hip flexion, back extension (again), and ankle dorsi flexion occur. The muscles in the back that cause the extension do so isometrically, which is why they don't cause motion.

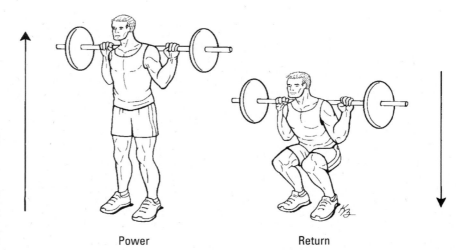

Figure 11-1:
Power-and-return phases of the squat exercise.

Power   Return

Illustration by Kathryn Born, MA

Some of the common flaws when doing a squat include the following:

✔ A poor base of support, in which the feet are placed either too close together or too far apart. Typically, they should be shoulder width apart.

✔ Folks come down too far during the return phase, putting undue stress on their hips and knees.

✔ During the power phase, participants begin to lean forward and rely on their backs to pull them upright again — a real killer on those muscles! By looking straight ahead instead of at the ground, the athlete can help avoid the problem of leaning forward.

Checking out your gait (walking)

Unlike a squat, walking is a considerably complex task. Walking challenges every aspect of your balance because it involves continually shifting your weight and executing complex sequences of motions for each joint involved. It's no wonder that, as people age, they tend to fall more and more frequently.

In this example, we break the gait cycle into eight phases (although it can be broken into more or fewer, based on the preferences of the person doing the analysis). In the gait cycle, the phases include both legs, typically one in contact with the ground and the other in a swinging phase. As you can tell from Figure 11-2, as one leg completes a particular phase, that same phase begins on the other leg, a pattern that occurs over and over and over again as you walk, or ambulate.

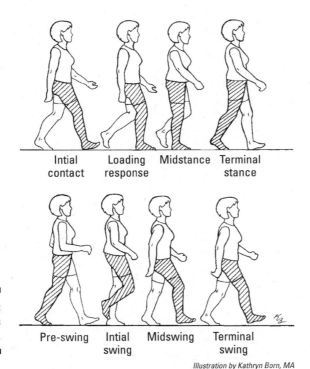

Figure 11-2: The phases of walking.

Intial contact   Loading response   Midstance   Terminal stance

Pre-swing   Intial swing   Midswing   Terminal swing

Illustration by Kathryn Born, MA

Here's what's happening during the different phases:

> ✓ **Initial contact:** The heel of one leg makes contact with the ground. To achieve initial contact, the leg needs to get to the point where the heel can land. For this to occur, the hip flexes and pulls the leg forward (an action that is discussed further in the swing) as the knee extends

and the ankle is dorsiflexed in an attempt to get the heel ready to hit. Further up the chain, take a look at the torso and arms. You can see that the arm opposite the striking heel is swinging forward while the torso rotates toward that front foot. *Note:* Some call this motion the *heel strike.* We use the term *initial contact* because more than just the heel is affected here.

In a *dorsiflexion,* or *dorsiflex,* your toes get closer to your shin, even though you aren't lifting your toes up.

- **Loading response:** After the heel hits the ground, the knee and hip flex to absorb the weight of the body. As the weight is absorbed by this process, it decreases the amount of force that is transferred down through the heel. At this point, the torso continues to turn a bit, and the opposite arms are still in front of the body and begin to move back (posterior).

- **Midstance:** The body, or center of mass, is located directly over the foot on the ground. At this point, the ankle continues to dorsiflex as the body moves forward. The knee moves toward more extension, as does the hip. At this point, the torso becomes more neutral and returns to its beginning position as the arm continues to move farther back. This is when balance really becomes important. All your weight is on one leg. Because your legs are side by side, you can see how, during this phase, you're likely to move side to side a bit as you also try to move forward.

- **Terminal stance:** As you continue to move forward, your heel begins to lift off the ground. This happens because, as dorsiflexion progresses, it causes the heel to rise. When the toes are really the only things on the ground, the knee initially flexes and then moves into extension as the push-off (pre-swing) comes into play. The hip continues to extend back and also plays a role in propelling you forward during the push-off. At this point, you remain upright, as your opposite leg gets ready to move into the initial contact or heel strike phase for that side.

- **Pre-swing:** The pre-swing phase is pretty much the same as the terminal stance, but the emphasis is different. The goal of the terminal stance phase is to achieve a position in which the toes are the only thing on the ground. In the pre-swing phase, the focus is on the act of propelling you forward. To achieve transition to the next phase, push-off must occur, and it is achieved with *plantar flexion* at the ankle (which points the ankle and toes to the ground), extension at the knee and hip, and, once again, an upright posture at the torso and head. Some also refer to this phase as *push-off* or *propulsion.*

- **Initial swing:** The pre-swing's effect on propelling the person forward demands that the foot be picked up higher so that it doesn't drag or catch on the ground. In the initial swing phase, the ankle is dorsiflexed (lifting the ankle and foot up) to clear the ground and not trip. Meanwhile, the knee is flexed, as is the hip; a lag in either of these motions causes you to catch your toe and trip. At this point, the same-side shoulder is flexed but moving back, and the opposite side arm is back and, like the leg, is moving forward.

✔ **Midswing:** The continuing forward motion of the leg makes up the mid-swing. In this phase, the hip continues to flex while the knee begins to extend and prep for the terminal phases and heel strike. The same-side arm continues to move back while the opposite arm moves forward; the torso rotates toward the leg that is up. The ankle is still dorsiflexed to make sure the foot is clearing the ground.

✔ **Terminal swing:** At the terminal swing phase, the leg slows down its forward momentum a little to achieve an optimal stride length. If the terminal swing phase doesn't control momentum of the leg moving forward as it should, you overstep. The hip continues to flex, the knee extends to maximize the stride without overstretching, and the ankle begins to plantar flex. Stepping with too short of a stride causes the impending initial contact to occur too early, and you won't land on your heel as you should.

The surface you're walking has effects on each of these phases and the extent to which some of the motions occur. For example, if you're walking on a sandy beach, you have to push a lot harder in the terminal stance and pre-swing phases, but the loading response creates less knee and hip flexion because soft ground has fewer forces.

The key difference between walking and running is that, when running, at some point, the whole body is off the ground, usually referred to as *flight*. Additionally, you usually don't see the heel hitting the ground when you run; instead, the contact almost exclusively occurs at the front of the foot and toes.

As complex as walking is, you don't really don't think about it much. Pretty cool, huh? Chalk that up to your amazing body!

Observing a kick in action

Another complex developmental task most kids use growing up is the kick. A kick involves integrating movements between both upper and lower extremities, as well as between left and right limbs, and requires precision upon execution. Heck, who doesn't want to score a goal? Timing and sequencing of each phase is integral.

To keep things relatively simple, we use a penalty kick approach as the example (see Figure 11-3). With a penalty kick, the athlete doesn't have to negotiate around a defender, moving targets, and so on. The logical way to break this task up is into the preparation (approach), execution (kick), and follow-through phases.

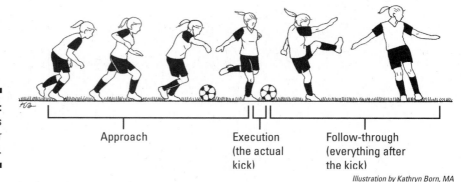

Figure 11-3:
The phases
of a soccer
kick.

Approach

Execution
(the actual
kick)

Follow-through
(everything after
the kick)

Illustration by Kathryn Born, MA

The approach

The approach, illustrated by the first four motions shown in Figure 11-3, includes movements from stepping up to the ball to planting next to the ball in preparation for the strike (when you put one foot next to the ball and begin to kick it with the other one).

Refer to the preceding section, "Checking out your gait (walking)," to understand the movements of the knee, hip, and ankle during this phase. Where the approach begins to differ is in the way the you begin to lean forward and rotate your body as you move toward planting next to the ball.

When you plant, you have to maintain your balance on the one side. In this case, the actions are largely isometric in the planted leg. The striking (kicking) leg begins to draw back to wind up by flexing at the knee and extending at the hip as the torso rotates to the side of the planted foot. The torso's leaning forward facilitates the explosiveness of the strike and is an important aspect of this phase.

The execution

In the execution phase (which happens between the fourth and fifth image in Figure 11-3), the kicking side knee extends and the hip flexes as the ankle plantar flexes until it makes contact with the ball. While the leg is being brought forward, the torso begins to rotate really quickly toward the striking leg, a movement that helps produce a more powerful kick. The arms follow the torso and rotate, and the non-kicking side shoulder flexes and *adducts* (comes toward the middle of the body) horizontally while the kicking leg shoulder *abducts* (moves away from the body) horizontally and extends, both with extended elbows.

An interesting side note is that just before contact, the foot appears fully plantar flexed, but upon making contact with the ball, it is pushed even further into this position because of the force of the contact.

The follow-through

After striking the ball, the torso recoils back to its original position, in essence rotating back away from the kicking leg side, and the arms return to their original starting position, as the final image in Figure 11-3 shows. The kicking leg then returns back to the ground as the hip and knee on the kicking side extend and the kicking ankle dorsiflexes to return to its original position. The non-kicking side pretty much remains the same and continues to stabilize and balance the body.

Analyzing phases in throwing

Considered a survival skill, throwing has various developmental patterns. Although the throwing action can happen in a number of different ways (shot put, javelin, and so on), here we explore the baseball pitch because it's a good example of a complex task that is highly dependent on sequencing and timing. Pitching a baseball involves and relies heavily on the coordination between the lower and upper extremities, as well as coordination between both sides of the body. While the throwing side does one thing, the glove hand does another.

Even though baseball pitchers often pitch in two different types of motions (the windup and the stretch), we limit the discussion here to the throwing motion related to the windup mechanism. In this example, we break the pitch up into six separate phases (see Figure 11-4): windup, early cocking, late cocking, acceleration, deceleration, and follow-through.

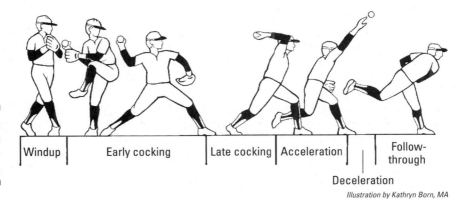

Figure 11-4: The phases of a base-ball pitch.

| Windup | Early cocking | Late cocking | Acceleration | Follow-through |

Deceleration

Illustration by Kathryn Born, MA

The windup

Despite being the namesake of the entire process, the windup can also be one of the individual phases within this analysis technique. This particular portion of the pitch sequence includes all movement that occurs up until the ball is removed from the glove.

The pitcher gets to the top of the pitching mound and stands on the rubber. At this point, everything starts. Here's what happens during this phase:

1. **Standing on the rubber, the pitcher begins to step backward and then rotates and brings his leg up in preparation to step toward the plate and throw.**

2. **To achieve the upright position, the player flexes his hip and knee while his ankle on the stepping, or *lead,* leg (the one in front) typically plantar flexes. At the same time, the abdominal muscles, the gluteal muscles (buttocks), and the back all work to maintain a balanced position.**

   In this stage, the glove hand and ball hand are in the mitt with the elbows flexed and shoulders at the sides.

3. **The pitcher begins to push off the rubber, with the planting foot (the foot on the same side as the throwing arm) toward the plate and batter.**

   To push off, the planting ankle dorsiflexes initially and then quickly changes over to plantar flexion while the knee and hip both extend on that side. As with the gait cycle, the leg that was initially raised is being driven toward home plate and extended at the knee and hip while the torso begins to rotate. Additionally, the hip is abducted in an attempt to make the necessary lateral move.

Early cocking

When the ball is removed from the glove, the early cocking phase begins. The movements between the upper and lower extremities must be integrated during this phase:

- **Lower extremity:** The front leg continues to move toward the plate while the back (push-off) leg propels the body forward. In the front leg, the ankle is plantar flexing, the knee is extending, and the hip is both abducting (moving away from the body to the side) and externally rotating. In the push-off leg, the knee and hip are both extending as the ankle plantar flexes. One of the keys to this phase is that the abdominals must contract, effectively rotating the torso to the side of the body opposite from the ball.

- **Upper extremity:** In the throwing arm, the shoulder joint is being both horizontally abducted and externally rotated. The elbow moves from a flexed position to an extended position. On the glove side, the shoulder flexes and rotates externally, and the elbow extends as the pitcher prepares to move the glove toward the target.

Late cocking

Just as with the early-cocking phase, both upper and lower extremities are involved in the late-cocking phase. Remember, throughout this phase, the ball is still in the hand:

- **Lower extremity:** The center of mass (torso) has moved forward significantly, and the front leg begins to assume a lot of the body's weight. The foot is in full contact with the ground, and the ankle is more dorsiflexed than previously. The knee flexes to absorb the body weight, and the hip flexes and rotates internally. Continuing its movement from the early-cocking phase, the torso continues to rotate, effectively transferring energy from the lower extremity to the upper extremity.

- **Upper extremity:** The arm moves backward, even as the lower extremity moves forward a good bit. To generate velocity, the torso twists. The throwing side shoulder joint (the glenohumeral joint) continues to rotate externally and abduct horizontally while the elbow begins to flex as the body moves forward. The glove arm now is being drawn back as the shoulder extends and the elbow flexes.

This phase is important for three reasons: One, this is the primary phase where energy is transitioned between the lower and upper extremities. Two, in this phase, the shoulder is asked to abruptly go from external rotation to internal rotation in preparation for the acceleration phase. Three, how the athlete performs during this phase of the pitch is integral to avoiding shoulder injury and instability.

Acceleration

Acceleration is characterized by the ball finally moving forward until it leaves the hand. Again, both lower and upper extremities pay key roles in this phase of the pitch:

- **Lower extremity:** Through knee flexion, ankle dorsiflexion, and hip flexion and internal rotation, the front leg continues to absorb the weight of the body as the center of mass moves forward. Meanwhile, the back leg finally pushes off the rubber by extending the hip and knee and plantar flexing the ankle. After the initial push-off, the knee flexes and the hip continues to extend until the deceleration begins to happen.

- **Upper extremity:** During this phase, the ball moves forward from behind the body. The throwing arm adducts horizontally and rotates internally to propel the ball forward. In addition, the elbow of the throwing arm extends while the torso continues to rotate to the opposite side. The glove arm essentially isn't doing much besides tucking itself toward the side of the torso and holding still.

Believe it or not, this phase isn't the one that has the most muscle activity. That accolade goes to the deceleration phase (explained next). What does happen of note here is that the shoulder rotates the fastest during this phase.

Deceleration and follow-through

Although typically referred to as two separate phases, the motions of deceleration and follow-through are pretty much the same, with a key difference: The deceleration phase includes a very distinct difference in muscle activity.

The arm needs to slow down at this point in the pitch; therefore, the muscles fire a lot to brake (slow) the internal rotation and horizontal adduction that's occurring. This braking occurs in the deceleration phase, whereas the follow-through is a continuation of the motion, but the muscle activity quiets down tremendously. In fact, the deceleration phase is where most muscle activity occurs.

Here's what happens in both the lower and upper extremities:

- **Lower extremity:** As the body continues to move forward, the lead leg still accepts the weight through knee flexion, ankle dorsiflexion, and hip internal rotation and flexion. In addition, in this phase specifically, the torso begins to flex forward in an attempt to absorb and help slow down the body after ball release. The back leg now begins to rise up behind the body by extending the hip and flexing the knee as the body leans forward.

- **Upper extremity:** Long after the ball has been released, these two phases result in the throwing shoulder continuing to adduct horizontally and rotate internally, while the elbow finishes in flexion. The glove hand sustains its position next to the torso, and the phase ends with the player assuming a position where he is ready to field the ball.

Common pitching flaws

Some major flaws that can be identified during this activity include

- **The lack of balance during the wind-up:** If you aren't able to maintain balance during wind-up, you begin to lean, throwing everything else off.

- **Pitchers not initiating their hips early enough:** This problem commonly occurs in the early- and late-cocking phases and leads to a decrease in external rotation. Often, throwers don't move their hips as they rotate. Instead, they try to accomplish all that rotation in their spine, which isn't designed for that.

- **Players dropping their elbows and nearly pushing the ball:** This issue occurs during the acceleration phase.

  Because this flaw can potentially be catastrophic, causing ligament damage, it needs to be addressed early.

- **Lack of follow-through:** Mitigating the forces created during the entire motion is important; otherwise, you end up relying on the little rotator cuff muscles to slow everything down. Proper body positioning and going through the necessary follow-through phase lets the arm slow down without putting undue stress on the limb.

Part IV
Mind-Body Connections

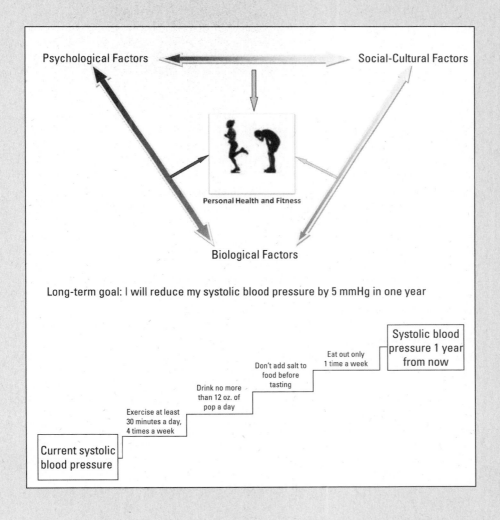

Psychological Factors

Social-Cultural Factors

Personal Health and Fitness

Biological Factors

Long-term goal: I will reduce my systolic blood pressure by 5 mmHg in one year

Systolic blood pressure 1 year from now

Eat out only 1 time a week

Don't add salt to food before tasting

Drink no more than 12 oz. of pop a day

Exercise at least 30 minutes a day, 4 times a week

Current systolic blood pressure

Check out how you can improve performance by getting your mind to help your body relax and "see" what it's supposed to do at www.dummies.com/extras/kinesiology.

In this part...

- Get acquainted with the components that make movement possible

- Understand how your body, your thoughts, and your environment work together to determine your overall health and fitness levels

- Build a stairway to success by setting challenging, but realistic, physical fitness goals

- Motivate yourself and others to engage in healthy physical activities and overcome common barriers to exercising

- Identify when exercise becomes an unhealthy compulsion and find ways to provide support and assistance to those who may be exercising too much

- Discover how physical activity can make people feel better

Chapter 12

Staying Connected: The Bio-Psycho-Social Model

. . . .

In This Chapter

▶ Understanding how complex bio-psycho-social interactions impact fitness and health

▶ Examining how biology, psychology, and culture can make you sick — or healthy

▶ Using the bio-psycho-social model when studying kinesiology

. . . .

*I*t's estimated that nearly 108 trillion people have lived on Earth. No two (even no two so-called "identical" twins) have ever been exactly the same. Each of us is born with a unique biological makeup, develops a unique psychology character, and has a unique set of interpersonal and social experiences. As a result, we become one-of-a-kind creations, with our approximately 20,000 genes continuously intermingling with our millions of individual thoughts and feelings in an infinite number of environmental experiences.

The billions of ways all these biological, psychological, and socio-cultural pieces blend together and affect one another is what makes us who we are and influences what we do and what we become. Whenever you want to understand the health and fitness of individuals and groups, you need to consider these factors. That's where the bio-psycho-social model comes in. It provides a construct through which you can better understand the reasons why some people get sick and others don't, for example, and how you can help people engage in behaviors that allow them to be fit and healthy for a lifetime.

In this chapter, we introduce the bio-psycho-social model and explain its usefulness and function. And because it is a major concern to exercise scientists and medical professionals around the world, we use diabetes as an example to show how every person's unique biological, psychological, and social environment often work together to cause, postpone, or prevent health problems.

Introducing the Bio-Psycho-Social Model

People who study the intricate workings of the human body for a living — including exercise scientists and kinesiologists — try to get a grip on the very complicated movements, functions, and disorders of the body by reducing them to their smallest observable parts. They focus their attention almost entirely on microscopic events and rely heavily on the language of (bio)medicine, (bio)chemistry, and (bio)mechanics to explain how the body works. In fact, you're holding a perfect example of this kind of reductionist thinking in your hands right now.

By simply scanning the pages of this book, you quickly see that most of our discussions, at least to this point, explore complex physical movements, such as lifting weights, throwing a baseball, and running a marathon, by examining the microscopic chemical and mechanical events that take place in the teeniest, tiniest parts of the muscles, nerves, bones, and blood cells. We describe, in great detail, how movements are controlled by interconnected networks of neurons chemically "talking" with one another (Chapter 3); how the air you breathe and the food you eat are chemically converted into energy (Chapters 4 and 5); and how the muscles, bones, tendons, and ligaments work like little gears, levers, and pulleys to get your body — or at least some of its parts — to go from one place to another (Chapters 7 through 10).

But there's more to movement than moving. In the next sections, we explain how the bio-psycho-social model, shown in Figure 12-1, helps you identify all the factors that impact movement and explain why budding kinesiologists need to pay as much attention to those factors as to the physiology of movement.

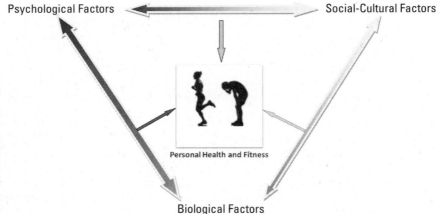

Figure 12-1:
The bio-psy-cho-social model of personal health and fitness.

Psychological Factors

Social-Cultural Factors

Personal Health and Fitness

Biological Factors

Illustration by Wiley, Composition Services Graphics

Understanding the model

Sure, a really good understanding of the biological and physiological details involved in moving is essential, but how much do meticulous descriptions of nerve signals, muscle contractions, oxygen consumption, and energy production actually reveal about true human movement? True, they explain some really important things about *how* your body is able to propel you from point A to point B, but they don't say anything at all about *why* you even want (or in many cases, don't want) to move in the first place. Nor do they reveal *where* you may go after you decide to move, *what* types of movements you'll choose to make, *when* you'll make them, or with *whom* you want to be when you do.

The five W's (who, what, where, when, and why) of movement are equally — if not more — important to consider than the mere "how" questions (how do the muscles move, how does oxygen fuel the cells, and so on) you answer when you focus solely on biological processes.

 Some of the many factors that interact with one another to determine a person's health, fitness, and performance level are the biological, psychological, and social factors, some of which we list in Table 12-1. Keep in mind that these are just a tiny fraction of the thousands of unique bio-psycho-social factors that combine to make you who you really are.

Table 12-1	Biological, Psychological, and Social Factors Impacting Health, Fitness, and Performance	
Biological Factors	*Psychological Factors*	*Social Factors*
Age	Anxiety or depression	Social support
Gender	Exercise self-efficacy	Instruction and coaching
Muscle-fiber ratio	Self-esteem	Social reinforcements
Agility and coordination	Motivation	Availability of child care
Illness or injury	Competitiveness	Health education
Reaction times	Goals and expectations	Fitness role models
	Coping strategies	Media influences
	Self-presentation needs	Cultural identification
	Health related habits	Access to fitness activities
	Resiliency	Societal expectations
	Prior fitness experiences	Healthcare availability
		Community funding

Seeing the big picture

Suppose you visit your local fitness center and peek in on a highly structured group spinning class. You can reduce what's taking place in this class to the cellular level and spend hours (if not days or months) describing in detail how each exerciser's body turns oxygen and nutrients into energy; how the oxygen moves through the lungs to the blood and onto each cell of the body; what precise chemical and mechanical changes take place in the bikers' muscles with each rotation of the pedals; and so on.

As important as these things are to know, if you focus solely on the biology of moving, you may end up missing the big picture. Questions that remain unanswered with this sort of nuts-and-bolts approach include the following: Why did these folks come to this particular fitness center today? For that matter, why do they engage in exercise at all? Why are these people here when hundreds of others who live nearby are not? Why have they decided to go to this spinning class instead of swimming laps by themselves or attending the Zumba or yoga class next door?

Only by considering your movements as an integration of your biological nature, your psychological makeup, and your social and cultural environment can you truly begin to realize what studying kinesiology means. After all, kinesiologists and their clients, students, patients, and athletes aren't machines or simple bundles of muscles, blood, and bones held together inside big bags of skin. They're unique, thinking, feeling human beings. So everything you do — or don't do (including the way you engage in physical activity) — must be considered from a broad bio-psycho-social perspective.

Taking a Look at One Sweet Bio-Psycho-Social Example: Diabetes

As we explain in our discussion of metabolism in Chapter 4, life itself depends, in large part, on the vast amount of sugar (glucose) that is always floating around in the blood. No sugar, no energy; no energy, no life. It's as simple as that.

If anything keeps blood sugar from getting into the cell — its "place of business" — you'll soon find yourself in really big trouble because you'll have developed a life-threatening medical condition called *diabetes mellitus,* often shortened to just *diabetes.*

The type of diabetes — type 1 or type 2 — describes what happens to stop the sugar in your blood from doing what it needs to do. In the following sections, we briefly explain the types, causes, and treatments for diabetes, and then show the multiple factors that impact the onset and course of the disease to illustrate the bio-psycho-social model in action.

Explaining what diabetes is and does

Diabetes is categorized into type 1 and type 2 diabetes (there are a couple other, very rare forms as well). In both types, blood sugar (glucose) has trouble getting inside the cells. In the cells, glucose's job is to start the entire energy-producing cycle of life, as we explain in the discussion of the Krebs cycle in Chapter 4). The reason why glucose can't find its way into the cell, however, differs between the two forms of diabetes, as the next sections explain.

Understanding type 1 diabetes

Insulin is the hormone that is produced in the pancreas and allows glucose into the cells. In type 1 diabetes, either there is no insulin in the blood at all, or there simply isn't enough insulin to open the cell doors to let in an adequate amount of glucose.

The reason why the pancreas doesn't produce insulin (or doesn't produce enough) is, in part, a genetic problem — something you're simply born with — so type 1 diabetes is often noticed very early in life (which is why it was once called *early onset* or *juvenile* diabetes).

To get a better sense of what's going on with type 1 diabetes, think of the sugar in the blood as factory workers who show up for work at the energy plant (the cell), just like they're supposed to. Here's what happens:

1. The workers (the glucose) arrive at the factory (the cell) ready to do their job only to find the doors locked and no security guards (insulin) around to unlock them. Because it can't get into the building, the glucose just sort of mills around outside the cell, waiting to see whether someone will eventually come by to open the doors.

2. The plant manager (a glucose sensor in the pancreas) doesn't really know what's going on but quickly notices that the workers can't get in and sends out an emergency call for more security guards to let them in. Unfortunately, these reinforcements never show up.

3. Soon the next shift of workers arrive, but they can't find anyone to let them in either. The essential work of the factory soon grinds to a halt, and all the workers (glucose) are left just hanging around outside.

Injecting insulin is akin to hiring a bunch of temporary security guards to stand outside the doors of the factory to let the glucose into the cell so it can go to work. Because type 1 diabetes results from a lack of sufficient insulin in the blood, it is sometimes called *insulin-deficient* or *insulin-dependent* diabetes.

Understanding type 2 diabetes

In type 2 diabetes, the problem isn't a lack of insulin to let the glucose in the cell; the body produces insulin in sufficient amounts. The problem is with the mechanism that opens the cell doors.

Going back to the factory analogy, plenty of security guards (insulin) are standing at the ready to help the workers (glucose) get into the factory (the cell), but something has messed up all the locks so that the keys the guards have can't open the doors. This is why type 2 diabetes is sometimes called *insulin-resistant* or *non-insulin-dependent* diabetes. Once again, the workers are locked out of the cells and begin to congregate in large numbers outside the factory walls. Without the sugar to do its life-sustaining work in the factory, body functions soon come to a standstill.

Because type 2 diabetes is a lot harder to understand, both in terms of its cause and treatment, it is a near perfect example of how biological, psychological, and societal factors impact the onset, course, and outcome of a disease.

Identifying the causes and treatments of diabetes

The cause and treatment of type 1 diabetes are pretty straightforward. Some flaw in the body's autoimmune system destroys the special cells (called *beta-cells* or β-cells) in the pancreas that make insulin. As a result, not enough insulin floats around in the blood to help the glucose get into the cells to start the energy-producing process. Case closed. Until scientists figure out a how to stop the body from killing its own beta-cells, the only option is to dump external insulin (from human or animal sources) directly into the blood.

In type 2 diabetes, however, usually more than enough insulin is circulating in the blood, but the glucose still can't get into the cells. Complicating matters is the fact that at one time everything seemed to be working just fine but then, for some unknown reason, the receptor sites (the locks on the doors) of the cell walls change in a way that disables the insulin's ability to unlock the doors to let the glucose in.

Because plenty of insulin is already available, simply dumping more into the blood (the treatment for type 1 diabetes) doesn't help. Instead, the challenge is to figure out why these seemingly healthy cells "decided" to change their locks, to stop them from doing so in the first place, and, if they do change, to fix them so that they work again — exactly the kind of research kinesiologists all over the world are working on every day.

Beginning with biology: It's in your genes

When a person develops any form of diabetes, it's at least partly because of his or her biological makeup — that is, the way the person is put together genetically. If someone's biology hasn't already set the stage for diabetes to develop, it probably won't.

The reason scientists know that both type 1 and type 2 diabetes are, at least to some degree, caused by genetic factors is by studying whether diabetes tends to run in the family. The question they're actually interested in boils down to this: If one person in a family has diabetes, do other members of that family, who have a similar genetic makeup, have a greater likelihood of developing the same disease? The thinking here is that, if diabetes is a matter of biology, the disease should tend to cluster within families (who are genetically similar to one another) — and it does.

Running in families

Whether you develop diabetes is, at least to a certain extent, related to who you are related to. Those who have a close blood relative with diabetes have a far greater chance of eventually acquiring the same life-threatening condition. Although the hereditary risks are somewhat different for type 1 and type 2 diabetes, both are directly linked to a biologically inherited vulnerability to the disease:

- **Type 1 diabetes:** If a member of your immediate biological family (father, mother, sister, or brother) is diagnosed as having type 1 diabetes, the odds are you're about 10 to 20 times more likely to acquire this condition than someone who does not have a close relative with the disease. If one identical twin has type 1 diabetes, the other twin will also have it nearly half the time.

- **Type 2 diabetes:** Somewhere between 60 percent to 90 percent of twins "match" one another, with both siblings either developing or not developing diabetes. This match rate is significantly higher among identical twins than in non-identical twins.

TECHNICAL STUFF

Aren't those monozygotic twins so cute!

When a woman gives birth to two babies during the same pregnancy, the babies are commonly described as either identical or non-identical (fraternal) twins. Generally speaking, people make this distinction on the basis of how similar the two babies are to one another. If they appear to be identical (same sex, body type, facial features, coloring, and so on), that's what they're called. For most purposes, that's close enough. but when scientists are interested in a person's true biological makeup, they get a lot pickier about the use of the term *identical.*

To scientists, being identical goes way beyond just looking similar to a sibling who happened to have been born at the same time; the twins must truly be made of the exact same genetic material. To be genetically identical, twins must have started out together as one, single, solitary fertilized egg that later separated into two identical eggs. Twins who develop separately from a single fertilized egg are called *monozygotic twins* (*mono* = "one" and *zygote* = "fertilized egg").

Because both twins are the product of the exact same sperm cell and exact same egg cell (the only places genes come from), they must be genetically identical. Non-identical twins, on the other hand, are the result of two completely separate egg cells that have been fertilized by two completely separate sperm cells at roughly the same time. These *dizygotic* (the fancy word for "two fertilized eggs") twins are no more genetically alike than two siblings born at different times.

Of course, even knowing that a strong family connection to diabetes exists doesn't mean developing the disease is all a matter of biology. As it turns out, biology is only the starting point.

Don't let the mere fact that diabetes runs in families fool you into believing that it's only family members' common biology that is responsible for this clustering of the disease. Families don't share only genetic material; they often share physical, social, and psychological environments. People who live, work, and play together tend to develop similar exercise attitudes and habits, health beliefs, and dietary preferences — a whole bunch of things that make it almost impossible to tease out exactly how much a given case of diabetes is due to a person's inherent nature (biology) and how much is due to factors in the social and psychological environment in which he or she was nurtured (raised). The upcoming sections delve into these other factors in more detail.

Paying attention to what you chew and do

With type 1 diabetes, lifestyle choices have relatively little to do with preventing or treating the disease. That's because this type of diabetes is the result of inheriting the specific set of genes that cause the body to seek out and destroy its own beta-cells in the pancreas.

Type 2 diabetes is very different. Although inheriting the set of genes that makes you particularly prone to the disease plays a part, whether you actually get the disease — and what you can do about it if you do — is largely up to you.

The two things that seem to have the most effect on whether you develop full-blown type 2 diabetes is what (and how much) you eat and whether (and how much) you exercise. It's this direct connection with diet and exercise that causes kinesiologists and other exercise specialists to be so interested in this particular form of the disease. Simply by helping people change what they chew and what they do, you can have a tremendous impact on one of the most debilitating, costly, and fastest growing epidemics in the world today.

Reducing fat to reduce diabetes

Obesity appears to be the single trait almost all (80 percent to 90 percent) people with type 2 diabetes share. Fat (technically called *adipose tissue*) causes the release of several hormones that have been shown to be related to insulin inefficiency. As a result, excess body fat, especially when it's concentrated in the abdomen, may decrease insulin's ability to help glucose get through the cell walls.

This knowledge is pretty exciting for medical and exercise specialists because it means that preventing and treating this terrible disease is easy. Reducing the fat you carry around and thus reducing your chances of developing type 2 diabetes is usually a simple matter (at least in theory) of eating less (and healthier) and exercising more. Consider these facts:

✔ **The amount and type of food you eat has a tremendous impact on body fat.** Eating a lot or eating processed foods and foods that contain large amounts of fat and sugar make maintaining a healthy body weight difficult. A diet based on fresh fruits, vegetables, and whole grains, on the other hand, helps you stay slim and trim.

✔ **Engaging in about 2½ hours of moderate exercise a week is enough to significantly reduce the likelihood of developing type 2 diabetes.** Many diabetes experts are convinced that one of the big reasons more and more (and younger and younger) people are developing type 2 diabetes is because of increasingly sedentary lifestyles. People just don't burn off the calories they eat, and the extras are often stored as fat.

TECHNICAL STUFF

Measuring fat

The amount of fat in a person's body can be measured in several ways. Some require fancy (and expensive) equipment; others use nothing more than a bathroom scales and a tape measure. Two of the more common higher end methods for measuring body fat are pretty accurate:

✔ **Bioelectrical impedance analysis:** With bioelectrical impedance, a small (and painless) electrical current is sent through the body. Because different types of body tissues (fat, muscle, and so on) contain different amounts of water, they slow down (impede) the electrical signal at different rates. Measuring the resistance the current encounters as it passes through the body gives a pretty good idea of that person's body's composition, including the percentage of fat it contains.

✔ **Hydrostatic weighing:** By knowing a person's weight on dry land and how much he or she weighs when totally immersed in water and then using some fancy formulas (originally set forth in Archimedes's "Law

of Buoyancy" we introduce in Chapter 2), technicians can calculate the density of the body and the percentage of fat it contains.

Of course, much easier ways to estimate body fat also are available. One is to use the Body Mass Index (BMI), which is simply an assessment of whether you are underweight, of healthy weight, overweight, or obese for your height. To calculate BMI, all you need to know is your height and weight and then plug those numbers into any of a dozen free BMI calculators available online.

A final way to estimate body fat turns out to be something you probably knew all along: Just look at the size of a person's belly! By measuring someone's waist (belly) and comparing that to the size of his or her hips, you get a waist-to-hip ratio (WHR) that is not only a quick and dirty estimate of overall body fat but is just about the only way to assess abdominal fat, which is even more closely associated with type 2 diabetes than excess fat distributed evenly throughout the body.

Noting how the mind also shapes the body

The level of obesity that puts people at risk for diabetes isn't solely a matter of their biological makeup. Although individual metabolism rates and certain diseases and disorders can contribute to fat storage, the decisions people make about what to eat and how often to engage in physical activity, to a far greater extent, cause them to gain weight. Simply put, if you're carrying around enough extra baggage to increase your chances of getting type 2 diabetes (and that may only mean a few pounds), the way you think and feel about food and physical activity — in other words your psychology — probably had an awful lot to do with it.

They don't call it "comfort food" for nothing

The way we feel — good or bad — has a lot to do with how much (and what) we eat and ultimately how much fat we put on our bodies. If you're feeling lonely, depressed, rejected, guilty, or just plain down and are like most people, you eat . . . and eat . . . and eat. To make matters worse, you tend to gravitate to foods that are really unhealthy. Be honest, how often do you crave a snack of plain carrots, celery, or asparagus when you're having a bad day? Chances are a really bad day is far more likely to send you searching for your sugary, salty, fatty, starchy favorites like ice cream, cookies, chips, candy, french fries, soda, and pasta.

In addition, we don't eat these fat-producing food only when we're down; we also eat comfort foods when we want to celebrate and have a good time: cake and ice cream for birthdays; wine, a rich meal, and a box of chocolates for anniversaries; a pizza party to end the sport season; celebratory drinks for successfully completing a big assignment; and so on.

The power of thoughts on our actions

How many times have you told yourself that you'd look better, feel better, and probably be a whole lot healthier if you would just get out and exercise a bit more? If you're like most people, you've probably had that little chat with yourself innumerable times. So if people understand how beneficial physical activity is, why don't they get off their couches or away from their computers and into the gym? In a word, motivation (or, in many cases, the lack of it).

Knowing what you *should* be doing to keep your body slim and trim is one thing; turning these grand and ambitious thoughts into weight-reducing actions is quite another. In Chapter 13, we discuss in detail the psychology involved in translating exercise-related intentions into exercise-specific actions. The thing to keep in mind now is that, when it comes to developing a healthy lifestyle, a positive attitude generally brings about positive results, and a negative attitude brings about negative results. Often what's in a person's head has a huge impact on what eventually shows up on that person's body.

Gauging the impact of your family circle

Many of the things people do that affect how much fat they carry around on their bodies — like what they eat and how much they exercise — are the direct result of where, when, and with whom, they grew up. Your environment exerts a number of subtle and not-so-subtle influences on how you behave. It can affect how you talk, dress, comb your hair, and so on. More importantly, it can also shape your body by influencing the way you think and feel about the food you eat and the attitudes you have about what constitutes proper exercise.

Individual food preferences are closely linked to culture and society. When you eat, where you eat, how you eat, and, most importantly, what you eat are greatly influenced by regional, ethnic, and cultural environments. Whether you're inclined to gobble up — or throw up — a large serving of *haggis* (pudding made from sheep livers, hearts, and lungs), *chitlins* (the small intestines of a pig), *head cheese* (meat jelly made from the head of a calf or pig), caviar (fish eggs), or *prairie oysters* (calf testicles) depends on what things your family and friends taught you were and weren't good things to eat.

Even the everyday decisions you make about whether to snack on a piece of fresh fruit or a handful of chips or whether to drink a glass of water instead of a sugary soda are based on the types of food available in your home, at your school, and in your community and what those around you taught you about the true meaning of "good eats."

Having a real opportunity to engage in physical activity often goes well beyond where people live, work, and play. Sometimes whether you get a chance to experience and appreciate the true benefits of physical activity has as much to do with your social support network as your physical environment. Getting fit and staying that way is a lot easier when you're surrounded by family and friends who supply you with not only positive role models but also the continuous encouragement each of us needs to stay active for a lifetime.

Being a product of your environment

You've probably heard the old saying, "When opportunity knocks, you need to open the door." But what if you live in a place where the opportunity to engage in physical activity rarely, if ever, knocks? That's what happens when

- ✔ Your community has more private golf clubs and fitness centers than convenient and attractive public parks and pools.

- ✔ Your neighborhood hasn't invested in sidewalks and bike paths to encourage kids and adults to travel from place to place by burning calories instead of fossil fuels.

✔ Your school district decides to cut physical education from its curriculum in order to save money or to squeeze in more time for reading, 'riting, and 'rithmetic.

✔ Your area is so unsafe that parents would rather their kids stay indoors playing football or soccer on a computer than going outside to enjoy the games themselves on the local playground.

✔ You don't learn to value sports and physical activity simply because society, in general, thinks having such opportunity is not important for someone of your race, gender, body type, ability, or age.

Noting society's role in sickness and health

As we explain in Chapter 3, the nerves responsible for thoughts and emotions certainly have a physical structure. Only through the intricate workings of thousands of cell bodies, axons, dendrites, myelin sheaths, and neurotransmitters can messages be sent from one nerve to another. Yet, even though it's your body that ends up doing the things that make you healthy or sick, what you choose to do is usually the result of what's going on in your external environment.

Society can exert powerful influences on individual thoughts and feelings — and ultimately personal behaviors and biology. Plenty of powerful socio-cultural factors often do more to promote illness rather than health. Consider the following examples, where societal factors play an important role in whether people engage (or not) in behaviors that have a tremendous impact on their health:

✔ **Societal influences on diet:** Just think about how the explosion of a fast-food culture that all-too-often relies on the sale of quick, predictable, inexpensive, high-fat, high-calorie, high-sodium, high-sugar meals has changed the way people in the U.S. — and many Western cultures — think about eating. Then think about how many commercials you've seen — even in the past 24 hours — for the food sold at these restaurants. Think about the ingredients that are now served in school cafeteria meals; then think about how school administrators contend that they have to offer these non-nutritious meals to children because the kids refuse to eat the healthy foods and the healthy foods are just too expensive to prepare and preserve.

✔ **Societal influences on physical activity:** Think about the way most of us have come to rely on cars rather than walking or biking to work or school; then think about how much money our community is spending on building roads and bridges compared to the construction of safe sidewalks and bicycle paths.

Up in smoke: Societal influences on smoking

Until the mid-1970s, virtually every influential institution in society teamed up to provide subtle and less-than-subtle incentives to get people to begin smoking at a young age and to continue the deadly behavior throughout their lifetimes. Consider these examples:

✔ **Government:** Beginning in the 1930s, the tobacco industry and tobacco growers started receiving millions of dollars a year in governmental subsidies to make growing tobacco and producing cigarettes profitable. These subsidies, combined with relatively low taxes on cigarettes and few restrictions on smoking and sales (unattended cigarette vending machines appeared in nearly every gas station, restaurant, hotel lobby, bar, and convenience store in the country) made cigarettes easily accessible and affordable; even kids could pick up a pack of cigarettes with ease.

✔ **Education:** Smoking cigarettes in public areas of colleges and universities — and even some high schools — was once the norm. Seeing professors, staff, and students lighting up in dorms, cafeterias, libraries, faculty offices, sporting venues, lecture halls, and classrooms was commonplace well into the 1970s.

✔ **Media:** Cigarette companies sponsored some of the most popular radio and television programs, even those whose audiences were largely composed of children. Cigarette commercials frequently used respected physicians, sophisticated movie stars, and idolized athletes to promote the safety and social acceptability of cigarettes. In many films, the main characters puff on cigarettes throughout the picture. (Go back and watch some old Humphrey Bogart films from the 1940s and 1950s, but keep in mind that the chain-smoking Bogie ended up dying of esophageal cancer at the age of 57.)

✔ **Business:** Until 1998, major commercial airlines allowed their passengers to smoke while traveling coast-to-coast. Even today, many other types of businesses such as bars, restaurants, and hotels fight to limit — if not completely kill — laws restricting smoking in their establishments. Smoking is big business.

A reversal has taken place in regard to societal attitudes toward smoking. Since the mid-1970s, virtually every part of society has retreated from promoting and enabling smoking. Major policy shifts in government, education, media, and business have made smoking one of the most inconvenient, costly, and socially condemned public behaviors.

The result has been dramatic decreases in both smoking rates and the incidence of smoking-related diseases. Consider, for example, that between 1940 and 1975, the percentage of U.S. adults who were classified as smokers remained relatively unchanged (roughly 40 percent). As more restrictive antismoking policies began to take hold and people's perceptions regarding the health risks, the economic costs, and the social disapproval attached to smoking changed, the percentage of smokers was cut in half (less than 19 percent) by 2012.

✔ **Societal influences on smoking:** The socio-cultural influences that initially promoted, but later reduced, cigarette smoking is a classic example of the way bio-psycho-social factors interact with one another to impact individual health and wellness. Once considered — and portrayed as — a sign of sophistication, sexiness, and rugged individualism, smoking is now viewed not only as unhealthy (which it has always been) but also as a socially unacceptable behavior. The sidebar "Up in smoke: Societal influences on smoking" has the details of this transformation.

Recognizing That Biology Is Not Destiny

At the beginning of the 20th century, when Sigmund Freud boldly proclaimed that "anatomy (biology) is destiny," he wouldn't have gotten much of an argument. Relatively permanent physical differences like gender, race, and age, as well as physical, mental, and emotional challenges, were assumed to pretty much define who we are and what we could (or even should) do. In those days little was known about the effects of psychology and sociology on behavior. Yet in the past century or so, we've come to understand a whole lot more about the factors that truly shape our destiny, and biology turns out to be just one of the many important elements.

A French social scientist by the name of Émile Durkheim (1858–1917), whom many still consider the father of sociology, was probably the guy most responsible for getting people to consider that they are more likely to be products of their environments than prisoners of their biology. He even popularized a French word, *milieu* (pronounced "mill-you"), that is commonly used to indicate the many ways the external environment can exert a tremendous influence on who and what a person becomes.

Operating within biological boundaries

To a certain extent, each person operates within his or her own unique biological boundaries. If, for example, the genes that you've inherited from your parents dictate that you're going to grow to a maximum adult height of 5 feet 3 inches (1.60 m), your dream of becoming the next LeBron James will probably turn out to be little more than that — a dream. (Of course, NBA star LeBron — regardless of how much he tries — has just about the same chance of winning a gold medal in men's gymnastics or figure skating or of winning the Kentucky Derby as a jockey sitting astride a race horse.)

So does this mean that physical stature alone determines capabilities? No. Consider that, although he would never be mistaken for "King James," Tyrone "Muggsy" Bogues stood exactly 5 feet 3 inches (1.60 m) tall and scored a total of 6,858 points while playing for four different teams during his 14-year NBA career.

Reaching your threshold: It's higher than you think!

Biology — whether it's your height, your reaction time, or your ratio of fast twitch to slow twitch muscle fibers (explained in Chapter 10) — tends to set only the uppermost boundaries of what can be accomplished. What most people don't realize is just how high these biological thresholds really are. Almost all of us sell ourselves short and fail to accomplish what we might have simply because we convince ourselves that our biology limits us.

Countless examples of what the human body and spirit are capable of illustrate that we can all push ourselves far closer to the limits nature has set for us. Here's is just a small sampling of famous athletes who have overcome physical (biological) limitations to perform at the highest levels of physical activity:

- **Jim Abbott:** A professional baseball player who pitched in the Major Leagues for ten years and even threw a no-hitter against the Cleveland Indians in 1993. He was named the nation's best amateur athlete in 1987 and won a gold medal in the Olympics in 1988. Abbott, by the way, was born without a right hand.

- **Robert (Rocky) Bleier:** Running back on four Super Bowl Champion teams for the Pittsburgh Steelers during his ten-year National Football League (NFL) career. Bleier played nine of those years after having part of his foot blown off by a grenade and being shot through the thigh while serving as an infantryman in Vietnam (the military even determined that he had a 40 percent physical disability).

- **Tom Dempsey:** A nine-year veteran placekicker in the NFL. In 1970, as a member of the New Orleans Saints, he kicked a 63-yard field goal against the Detroit Lions, setting an NFL record that stood for 43 years. Dempsey was born without any toes on his right (kicking) foot and no fingers on his right hand.

Even more remarkable are the thousands of Special Olympics and Paralympics athletes who push themselves daily to overcome biological challenges to achieve at the highest level of personal performance.

Using the Bio-Psycho-Social Model

It's a mistake to think that anything people do and ultimately the levels of fitness, performance, or health they attain can be explained by examining their biology, their psychology, or their social experiences in isolation. Although it can sometimes be very difficult, kinesiologists need to remind themselves that these three areas interact and impact one another.

Given the complex interactions among dozens (if not hundreds or even thousands) of biological, psychological, and socio-cultural factors that impact the onset (or avoidance) of type 2 diabetes, for example, you can easily see that the only way to contain this expanding global epidemic is to take a multi-faceted approach that involves researchers and clinicians who specialize in each of these three areas — biology, psychology, and sociology.

Diabetes is certainly not the only disease or disorder that results from such complex exchanges between our bodies, our brains, and our backgrounds. The susceptibility to a number of diseases and conditions — many forms of cancer; heart attacks and strokes; degenerative diseases; shingles; ulcers; depression and anxiety; asthma; and muscle, tendon, joint, and bone injuries; even acne, dandruff, cavities, and bad breath — is usually the result of each person's unique bio-psycho-social makeup.

Whether you ultimately end up healthy or ill, happy or sad, or fit or fat will be the result of who you are (your biology), what you think and feel (your psychology), and the influence of your physical, social, and cultural surroundings (your society). To fully understand kinesiology — the "science of movement" — you must first understand why people do the things they do. Staying connected with the lives of your clients, your patients, your athletes, and your students can be accomplished only if you constantly work from a bio-psycho-social model of health and wellness.

Chapter 13

Sticking with It: Motivation and Adherence

In This Chapter

▶ Recognizing how exercise and training plans get sabotaged

▶ Setting challenging but realistic short- and long-term goals

▶ Controlling (instead of being controlled by) your goals

▶ Providing yourself (and others) with motivating feedback

*P*hysical fitness experts may be missing the boat by trying (usually unsuccessfully) to get totally sedentary people to start exercising. You can talk with them until you're blue in the face about the benefits of being physically active, but at some point, you have to concede the truth in the old saying, "You can lead a horse to water, but you can't make it drink." All the coaxing, bribing, and threatening in the world won't get sedentary people moving one minute before they're good and ready.

Your attention may be better focused on keeping those who have already begun exercising motivated to continue. Almost everyone — even the most inactive among us — has started an exercise program at some point in the past only to abandon it. Imagine the tremendous impact you could have on these folks' lives by helping them remain motivated enough to continue being physically active! In this chapter, we offer common sense — but scientifically based — strategies and advice you can pass on to others or simply use yourself when trying to maintain fitness and exercise programs.

Adopting the Right Mindset: Banishing the Enemy Within

You want to hear something really amazing? More than half of the people who begin exercising call it quits within the first six months. Often these people are written off as quitters or weak-willed losers who simply don't have the strength of character or personal resolve to stick to something that takes a little time and effort. But let's be totally honest: We've all been there at one time or another — making resolutions to join the local fitness club, to buy and routinely use a new piece of exercise equipment, or to spend a half an hour or so every day doing sit-ups. Despite all our good intentions, we give up. Why is following through on such worthy intensions so hard?

One reason sticking to a training and exercise program is so difficult is because people unintentionally sabotage themselves before they can experience the results they're looking for. Sometimes they quit because they haven't thought about how they're going to deal with the inevitable obstacles and challenges that threaten to interrupt long-term exercise programs. Other times they quit because they've set unrealistic or unattainable results.

Whatever the reason, failing to accomplish a goal quickly leads to frustration and discouragement. When disappointment and disillusionment set in, it's not long before the fitness club membership card is misplaced and eventually expires, that new piece of exercise equipment becomes little more than an expensive clothes rack in the corner of the bedroom, and the only sit-ups done are those that involve the assistance of the lever on a reclining chair.

To maintain the motivation necessary to continue physical activity long enough to reap the health, fitness, and appearance benefits they seek, people have to first stop undermining their chances for success.

Getting realistic

We all expect some kind of payoff from engaging in physical activity. After all, what would be the point of expending all that time, money, and energy if you expected absolutely nothing in return? Although looking forward to a time when increased activity will result in enhanced health, fitness, and physical appearance may be reasonable — and even motivating — expecting miracles is quite another thing.

The key is to manage your fitness expectations. Expect too little from involvement in physical activity, and you may never get started in the first place. Set unrealistic expectations, and you end up feeling frustrated and discouraged — sure signs that quitting is just around the corner.

To counter the self-defeating mindset that if it doesn't come soon or easily, it's not worth working for, remember this old saying any time you feel like throwing in the towel: "There are no shortcuts to any place worth going."

Identifying your true goals

One of the most frequent reasons exercisers give for engaging in physical activity is to lose weight. They usually even go so far as to specify the precise number of pounds they want to lose. Yet aside from your doctor (and maybe the carnival barker who guesses people's weight), who really knows what number pops up each time you step on the scales? Nobody. What everyone does know, however, is what you look like, how your clothes fit, and how much energy you seem to have. So why not focus on those things?

The faulty assumption behind the desire to lose weight is that appearance and health are directly associated to the number of pounds people carry around on their bodies. Sometimes that's true; sometimes it isn't.

 Even though you may tell yourself and others that you're interested in losing weight, that's probably not the case at all. Why? Because weight isn't the real issue. Ask yourself this question: Which weighs more, 15 pounds of lean, solid muscle or 10 pounds of wobbly blubber? Obviously, 15 pounds of anything weighs more than 10 pounds of something else. So which would you rather carry around on your body every day? Even though the muscle would cause you to weigh 5 pounds more, you probably would still choose the 15 pounds of muscle — which will likely make you look and feel better — over the fat. What people really want is to improve their health, fitness, and appearance, and merely losing weight does not always translate into achieving those goals.

 If you don't focus your attention specifically on what you really want to accomplish through physical activity, you can easily get distracted from what motivated you to begin exercising in the first place. Many people give up their workout routines because they don't see the dramatic weight loss they expect, even though they were improving their fitness, health, and appearance — the very things they truly wanted to accomplish in the first place.

 If you're really interested in sticking with an exercise or fitness routine, do yourself a favor and throw away your bathroom scales. Continuously checking your weight to see whether your exercise program is working tends to distract you from the real benefits of exercise and can sometime be so discouraging that you feel like giving up.

Taking care of the little things

Rather than obsessing about whether you've gained or lost a fraction of a pound since breakfast, focus your attention on practicing good behaviors rather than seeing good results. If you take care of the little things — engaging in physical activity that you enjoy and can stick with for a long time, like taking the stairs instead of the elevator, walking or riding a bike for short errands, snacking on a piece of fresh fruit instead of a handful of chips — the big things will eventually take care of themselves.

One good way to make sure you take care of the little things is to set a series of goals in important health-related aspects of your life. For example, establishing daily goals for nutrition ("I will eat at least one serving of fruits or vegetables with each meal), hydration ("I will drink eight ounces of water three times a day), sleep ("I will get at least eight hours of sleep every night"), and relaxation ("I will sit quietly and meditate for five minutes every morning before my shower") is a great way to supplement your exercise or training goals.

Creating the Right Kinds of Goals

Very few worthwhile achievements are accomplished by accident. Typically, they require a detailed road map that gets you from where you are now to where you want to be in the future. The mile markers on your exercise and performance road map are a series of increasingly challenging goals and objectives.

Sounds simple enough, but all goals are not equal. Although some goals can help you reach your dreams, most have little or no impact whatsoever, and a few make succeeding almost impossible. In this section, we outline the specific types of goals that can maximize your chances for success.

Seeing how ready you are to begin

How ready are you to make the move away from your unhealthy, sedentary lifestyle to a physically active, healthier set of behaviors? One simple way to make this determination is by using something called the *Transtheoretical Model (TTM)* of behavioral change. The TTM basically says that, as people adopt a new (healthier) behavior, they are in one of six stages of change, outlined in Table 13-1. Your challenge is to move (or to help others move) through these six stages.

Table 13-1	The Six Stages of Behavioral Change	
Stage	*Description*	*Mindset*
Stage 1 — Pre-contemplation	People in this stage aren't even open to the idea of engaging in a new, healthier behavior.	"I have no need or intention to exercise."

Stage	Description	Mindset
Stage 2 — Contemplation	Although they aren't actually doing much of anything to make themselves healthy, people in this stage are at least willing to entertain the possibility.	"You know, I really should become more physically active."
Stage 3 — Preparation	People in this stage are doing a few things that will eventually help them begin the healthy behavior.	"I just bought a new pair of exercise shoes and a membership at the local gym so I can start exercising more."
Stage 4 — Action	People in this stage have started engaging in the healthy behavior for the first time.	"I currently go to the gym and work out for 30 minutes at least three times a week."
Stage 5 — Maintenance	In this stage, people continue to engage in the healthy behavior, but they also recognize they need to do things that will keep them on the right path and avoid a possible relapse into their old (non-healthy) lifestyle.	"Two of my friends and I go out for coffee after each exercise session to talk about how we're doing."
Stage 6 — Termination	People in this stage begin to feel a complete psychological separation from their original (non-healthy) lifestyle.	"I feel and look so good now, I can't imagine ever stopping my exercise routine."

Getting SMART about your goals

A good goal, whether long- or short-term, is one that is meaningful, measureable, and attainable. Therefore, as you set your goals, make them SMART:

✔ **Specific:** Your goals need to be stated in very specific terms. Simply "trying" or "hoping" to do something is not a goal.

✔ **Measureable:** At some point, you'll need to evaluate your goals. Having a precise measurement will tell you if you've accomplished your goal or not.

✔ **Attainable:** Although goals need to be challenging, they also need to be realistic. Setting unattainable goals sets you up for frustration and failure.

✔ **Relevant:** Your goals need to be important to you. If your goals are meaningful *to you,* you'll find it easier to get (and stay) motivated.

✔ **Time-sensitive:** Your goals need to be met at some specific point in the future. This way, you'll know when to assess whether you've accomplished them or not.

Staying in control of your goals

For your exercise, activity, or performance goals to be truly useful, you must make certain your goals remain under your control. Even under the best of conditions, you'll occasionally be faced with events or actions by others that threaten your ability to stick with your exercise or training program. To the extent that you allow others to have control over your accomplishments, you have unconsciously given yourself a built-in excuse for failure.

Consider these examples — one from exercise, the other from sports — that illustrate how an inability to control your goals can lead to failure:

✔ **Relying on others:** Suppose your stated exercise-related goal is to attend the "hot" yoga class at the local fitness center four days per week. However, you live several miles from the fitness center and need a friend (who has a similar goal) to drive you to the class. So far, that's no problem. In fact, maybe the two of you can support each other's goals. But what if your friend gets sick, her car breaks down, or she just decides that she doesn't want to attend the class in the future? Obviously, you need to make different arrangements if you want to continue toward your goal. Unfortunately, what often happens is that you'll decide to skip the class for a day or two, which then becomes a week or two or even a month or two, and before you know if you've fallen out of your exercise routine and you're right back where you started. The point is, when you don't have complete control over your exercise goals, you put yourself at the mercy of others.

✔ **Setting a goal that is beyond your control:** Without a doubt, the most frequent goal set by athletes involves winning — winning the gold medal, winning the next game, or winning the league championship. Although at first blush these sound like perfectly reasonable goals for any athlete, if you look a bit closer, you realize that athletes really don't have absolute control over whether they win or lose. Their victory or defeat also has a lot to do with the skill level of their opponents, the performance of their teammates, the strategies and tactics selected by their coach, the officials' calls, or just dumb luck. None of these factors are actually under their control. These external events and people have as much impact on the eventual outcome of a game as the athletes' performance.

Persevering despite the unexpected

When studying kinesiology, you'll come across a number of laws, like Newton's three laws of motion (inertia, acceleration, and action/reaction) which we cover in Chapter 8. Well, here's another law governing fitness, wellness, health, and exercise: Murphy's Law.

Murphy's Law applies not only to kinesiology or exercise but also to all aspects of life. Simply, Murphy's Law states that "whatever can go wrong, probably will go wrong." And that's certainly as true when you're trying to stick to an exercise or training regimen as anywhere else.

Life happens, and it sometimes gets in the way of continuing your exercise or activity program. When your favorite piece of exercise equipment is broken or when you've suffered a minor injury, it's tempting to stop your workout routine. But that's a problem. Once you stop, even for a short time, you'll find getting back into your routine can be very hard.

Because the best laid plans of mice and men often go awry, even the goals we originally thought were under our control will sometimes be interrupted. To avoid being waylaid by Murphy's Law and remain active, you need to be willing to make adjustments. By modifying your activities somewhat, you won't have to quit entirely. If you have a sore ankle, for example, you may be tempted to forego your daily running routine until your ankle heals. Rather than quitting altogether, however, a better solution is to go to the pool and swim laps for the same amount of time you would normally spend running.

Mapping out long- and short-term goals

You set goals for two reasons: (1) to provide yourself with an ultimate destination, an accomplishment you want to achieve at some point in the future, and (2) to continually monitor your progress — and adjust your course if necessary — as you inch your way toward that long-term goal. Both your destination goal (long-term goal) and your daily monitoring goals (short-term goals) provide you with the motivation and information you require for success.

Aiming for your long-term goals

Long-term goals are essential because they point you in the direction of something you think is worth pursuing. Without this touchstone to continually refer to, you risk losing your way at some point in the long journey. As the great "philosopher" Yogi Berra once said, "You've got to be very careful if you don't know where you're going, because you might not get there." Truer words were never spoken. Setting long-term goals keeps you motivated because, with them, you always know exactly where you're going.

Following are examples of long-term goals for health and fitness:

✔ I will reduce (without medications) my systolic blood pressure (top number) by five points (5 mmHg) before my next annual physical exam.

✔ I will complete my first half-marathon by the end of next summer.

✔ I will reduce my waist measurement by one pant size within the next year.

Notice that these are all good long-term goals because they are totally under your control. They are also SMART (refer to the earlier section "Getting SMART about your goals" for details).

When you set your goals, never use the term *try*. Using the word *try* is akin to admitting that you don't have complete control over your goals and that you really wouldn't be surprised if you failed.

Setting short-term goals

Knowing where you want to go is certainly important, but it won't do much good if you still don't know how to get there, and that's where setting daily short-term goals comes in. Short-term goals provide you with a real-time map of every step you need to take on your own private road to success. With short-term goals, you know exactly what you have to do to get yourself one baby step closer to your ultimate goal.

By definition, long-term goals aren't realized quickly. As a result, keeping your "eyes on the prize" is difficult when you work day after day without experiencing a tangible sense of accomplishment. Too often you begin to focus on the enormity of the task that lies ahead and become overwhelmed and disappointed with your slow progress. Short-term goals help you maintain your motivation by redirecting your attention toward the small things you've accomplished rather than the big things you haven't — yet.

Suppose that your long-term goal is to reduce your systolic blood pressure by 5 mmHg before your next annual physical exam. To reach that goal, you need to set short-term goals like the following, each taking you one step closer to reaching your ultimate objective (see Figure 13-1):

✔ I will exercise a minimum of 30 minutes a day at least four times a week.

✔ I will not drink more than 12 ounces of soda pop in a day.

✔ I will not add salt to any of my food before I've tasted it and determined it truly needs it.

✔ I will not eat at a restaurant more than one time per week.

Long-term goal: I will reduce my systolic blood pressure by 5 mmHg in one year

Figure 13-1: Achieving your long-term goal, one step at a time.

Current systolic blood pressure

Exercise at least 30 minutes a day, 4 times a week

Drink no more than 12 oz. of pop a day

Don't add salt to food before tasting

Eat out only 1 time a week

Systolic blood pressure 1 year from now

Illustration by Wiley, Composition Services Graphics

Providing Useful Feedback to Others

Whenever you want to get others (or yourself) to begin or continue with a physical activity program, you're essentially in the motivation business.

The root of the word *motivation* is *motive,* which means "to provide with an incentive" and is itself based on the Old French word *motif,* meaning "causing to move." You can provide people with an incentive to start — or continue — moving in several ways:

✔ **Give them something they value or enjoy whenever they engage in the desired activity.** In other words, give them "carrots."

✔ **Coerce them into moving — often against their will — by giving them something they *don't* want if they fail to engage in the desired activity.** That is, give them the "stick."

✔ **Give them a little bit of reward for what they've done well along with a little instruction about what went wrong and what they need to do differently next time.** Essentially, you're combining the two preceding approaches and handing out "carrot-sticks."

Because behavioral scientists aren't too keen on using silly terms like *carrots* and *sticks,* they've come up with their own set of terms that pretty much mean the same thing. To them, *carrots* are *reinforcements, sticks* are *punishments,* and *carrot-sticks* are often called *feedback sandwiches,* which, despite appearances, is *not* a silly name because you're sandwiching information and instruction between other important pieces of performance-related information.

You can use each of these behavioral approaches as incentives to move someone in a desired direction. However, because punishments have been found to cause so many unintended consequences (resentment, anger, and

apathy, for example), the general consensus is that providing reinforcements and feedback sandwiches is more effective when you're trying to get someone to start or continue a physical activity program.

At some point, you'll be in a position to provide performance-related information to others as well as yourself. Coaches, teachers, athletic or personal trainers, fitness instructors, and physical or occupational therapists all need to know how to provide accurate, timely, and effective information to their students or clients. In this section, we tell you how to use a future-oriented, positive, and empirical approach when evaluating and providing feedback to others. By doing so, you can often enhance the performance of those with whom you work.

Giving positive reinforcements

To reinforce something simply means to make it stronger. When you reinforce a behavior, like going to the gym and working out, you're making that behavior stronger and more likely to be repeated in the future. One way to reinforce a desired behavior is to offer a small reward that will make this behavior more likely to reoccur.

Although treating yourself — or others — with external *(extrinsic)* rewards for a job well done is a very effective way to get yourself started, a time will come when you need to engage in the desired behavior for its own sake and not just to get some external prize. If you're motivated to work out only for external reasons, you may lose sight of the true internal rewards and benefits it provides. Then if the external rewards are ever removed, you have no reason to continue.

For reinforcement to work, the reward must be meaningful to the person receiving it. Suppose, for example, that you love a good, long soak in the tub. If you indulge in this special treat only as a reward for going to the gym, you're more likely to keep going to the gym on a regular basis so that you get to soak in the tub more often. For more detailed information on how to use rewards to increase or maintain the motivation to work out, head to the later section "Rewarding yourself for a job well done."

Giving feedback sandwiches to others

The most effective way to correct performance errors is to use a feedback sandwich, made of positive reinforcement, future-oriented instruction, and encouragement. In this sandwich, the "meat" of your message (constructive and specific feedback regarding what the athlete needs to do in the future) is sandwiched between two genuine, supportive statements.

To serve up a feedback sandwich, follow these three steps:

1. **Give honest, thoughtful reinforcement for something the person has done correctly.**

   Acknowledging the positive aspects of the athlete's or client's performance increases the likelihood that she will do those things correctly again next time. More importantly, when you begin with a positive statement, your client is more willing to listen to what you have to say. After you have her attention, you can slip in the meat of your message.

   Motor performances are usually made up of a chain of small actions. Most elements of any performance — even a "bad" performance — are actually done correctly. Because you want these correct parts of the performance to be repeated, you need to acknowledge them.

2. **Give future-oriented instruction.**

   Your clients don't need you to tell them when they've made an error. What they need is for you to tell them how to avoid making the same mistake in the future. The key here is to specify one simple change they should make next time.

   Because it's impossible for learners to change the past, help them focus on the one and only thing they can control — what they can do differently *next time*. Constantly harping about things they can't do anything about may make you feel better, but it only makes your learners frustrated, angry, and demoralized.

3. **Give general encouragement and support.**

   The final step in the feedback sandwich involves letting your clients know you have confidence in their ability to perform the skill properly.

   If you can't honestly say that you have confidence in their ability to perform the skill, you're setting them up for almost certain failure. If necessary, consider adjusting their responsibilities (and your expectations) so that you're asking them to perform only tasks they are truly capable of accomplishing.

The point of the feedback sandwich isn't simply to create a feel-good experience by buttering up or flattering your learner. You are deliberately pointing out what was done correctly so that it's likely to be repeated. If you focus too much on the errors and ignore what was initially done right, you may succeed in getting your learners to change what they did wrong but mess up something they did correctly the first time. Your goal is to remind them to repeat the good parts of the performance, even after they correct the errors.

Using physical activity as punishments — and why you shouldn't

Believe it or not, thousands of formally trained physical educators routinely and intentionally use the very thing they are professionally obligated to promote — physical activity — as punishment to correct disciplinary or performance errors their athletes or students make. In sports and in far too many physical education classes, using activity as punishment is as common as sweating.

We all pretty much know the drill by heart. If you're late for class, you run laps. If you miss a free throw, you run sprints. If you fumble the ball, you do push-ups. Unfortunately, after seeing so many coaches using as a punishment the very activities they should be encouraging, many of us mistakenly assume that this type of discipline is a necessary and natural part of sports. It's not, and it can have negative implications for lifelong fitness and health.

Nearly every day, newer and stronger scientific evidence shows that regular physical activity is essential for a long and healthy life. Largely in response to this call for exercise, physical education classes and athletic programs are sold to the public as a means of instilling young people with an appreciation and enjoyment of physical activity that can serve as a foundation for lifelong fitness. But imagine how unattractive the very thought of lifetime exercise is for someone who has been taught, through years of negative experiences, to despise and dread even the thought of physical conditioning.

Physical activity is a *good* thing. It is something to be encouraged and should never, *ever* be used to intimidate, threaten, or terrorize people into changing their behavior. After all, the role of the teacher or coach is to motivate, not bully, their students and athletes to a higher level of performance while providing them with a lifelong appreciation for exercise.

Practical Tricks and Tips for Sticking with a Physical Activity Program

Preconceived notions about physical activity can sometimes get in the way of starting or continuing an exercise or fitness program. That's why making small, almost imperceptible changes in the way you think and the way you structure your life — changes that will support your decision to adopt and maintain a healthier lifestyle — is so important.

Often, all you have to do is replace some of the old, negative views of exercise with positive ideas about how you can make physical activity fun, interesting, exciting, and enjoyable. In this section, we offer suggestions to help you focus on the positive and enjoyable aspects of exercise. Remember, long-term fitness is not about forcing yourself to do the things you hate. It's about finding and doing the things you truly enjoy.

Taking time, making time

One of the most often-used excuses for failing to adhere to an exercise and activity plan is that there are just not enough hours in a day to devote even half an hour to exercise. This adage summarizes what's really at stake: "Which fits into your busy schedule better: exercising for a half-hour a day or being dead 24 hours a day?"

Life is all about making choices. True, if you set aside a half-hour every day to engage in some form of physical activity, you have to give up something else. But because you can't add time to the day, you have to find a way to do a better job of using the time you have to accomplish the things you think are important. You may, for example, decide to watch one less television sitcom that day, send a couple fewer text messages to friends, or wait until tomorrow to post that new photo on your Facebook page.

Tailoring your activities to fit your needs

With physical activity, one size does not fit all. Selecting the physical activity that's right for you is absolutely essential if you have any intention of sticking with it for the long haul. Just make sure the activity you select does the following:

✔ **Is compatible with your lifestyle and interests:** Some people like to walk, run, or bike by themselves with only their thoughts for company. Others prefer to be led through a structured set of Zumba dance and aerobic routines by an energetic instructor. Still others prefer to play basketball or hockey with a group of friends.

✔ **Is compatible with your fitness and energy level:** Irrational exuberance at the beginning of an exercise program has turned a lot of people off from physical activity forever just after just a couple sessions. By failing to match their current level of fitness with their enthusiasm, they often end up with muscle aches and pains that convince them that the whole notion of becoming more physically active was really a bad idea right from the start.

✔ **Is something you can do for the long haul:** Fitness is a marathon, not a sprint, so make sure the activities you engage in are things you can imagine yourself doing for a long time. The health, fitness, and appearance benefits associated with physical activity will come, but only if you stick with it.

The frequency, duration, and intensity you're expending today or tomorrow isn't nearly as important as whether you're continuing to put forth that same kind of effort a year or two down the road.

Although individual preferences and fitness level dictate the physical activities that are best for you, another consideration is important. The degree to which you are currently comfortable with your physical appearance may determine whether you want to work out with a mixed gender group — or in a group at all. The good news is that even if you do feel a bit self-conscious about your appearance, there's nothing that says you have to be physically active with other people. You can work out in the privacy of your own basement or bedroom; you can go (fully clothed!) for a walk, a jog, or a bike ride by yourself. If you think about it for a while, you can come up with all sorts of things you can do by yourself to get in better shape. No one has to see you exercise — unless and until you want them to be seen.

Recognizing that doing something is better than doing nothing

Regardless of how motivated you are or how well you've planned your activities, on some days you're just not going to feel like exercising. It's natural, it's normal, it happens, so don't beat yourself up. At the same time, don't totally give in to this self-defeating impulse.

On these "I don't want to exercise" days, simply promise yourself that you'll engage in only a light workout; maybe you limit yourself to some gentle stretching or decide to work on only those aspects of your normal routine that you enjoy most. The point is to get yourself going. More often than not, after you get started — and you've released yourself from feeling compelled to work out — you'll naturally get into your total routine. If not, that's okay, too.

Mixing it up

The adage that "variety is the spice of life" is certainly true when it comes to spicing up your exercise and fitness life. Forcing yourself to engage in exactly the same physical activities over and over again, day in and day out, becomes pretty dull. And the one thing you never want your exercise routine to become is dull. Soon after you start thinking that exercise is boring, you begin seeing it as drudgery and a waste of time, too. And then guess what happens.

One of the best ways to combat boredom is to shake up your daily routines. Instead of biking outside every day, attend a spinning class once a week. Or maybe get completely crazy and go swimming, sign up for an aerobics class, lift weights, go jogging, or play a round of golf or a game of tennis with your friends. The change of scenery will do wonders for your attitude, and before you know it, you'll be looking forward to simply being active again.

Pushing yourself — but not over the edge

To benefit fully from physical activity, you can't give up at the first signs of fatigue or discomfort. Push through these mild to moderate feelings of unpleasantness to go beyond your current level of fitness. Exercise and conditioning is not always easy, but it shouldn't actually hurt, either.

A big challenge all active people face is distinguishing between feelings of temporary discomfort and full blown pain. Pushing yourself through the muscle burn that often accompanies a strenuous workout is one thing; pushing yourself when you're experiencing true muscle pain is quite another.

Don't make the mistake of thinking that, if a little exercise is good, a lot of exercise is better. When you push your body beyond a certain point, you're far more likely to experience injuries that impede your ability to continue your journey toward fitness and health. For a discussion of the specific problems associated with excessive exercise, head to Chapter 15.

Charting your progress

Sometimes people get discouraged and quit their training programs simply because they lose track of how much progress they've made. An easy way to remind yourself of how far you've come is to keep a daily exercise log. Just by recording duration, repetitions, and even how easy or difficult you found each workout, you have a running account of your accomplishments.

Few things are as reinforcing or motivating as progress toward a goal. By seeing exactly how your short-term goals are keeping you on track, you can give yourself a well-deserved pat on the back for everything you've accomplished thus far.

Rewarding yourself for a job well done

In addition to patting yourself on the back for a job well-done, you can also acknowledge your training achievements by giving yourself something a bit more tangible. Yes, we're talking about rewards. As we mention in the earlier section "Giving positive reinforcements," you can use rewards as a way to keep yourself (and others) motivated. Follow these steps:

1. **Think about what you would consider an extra special treat.**

   Make sure the reward is meaningful to you. Maybe it's a new piece of exercise clothing, the latest app for your smartphone, or dinner and a movie.

   You may want to consider rewarding yourself with that spiffy exercise outfit or a pair of the athletic shoes you've always coveted. That way you not only reinforce your effort, you look and feel better as you continue to exercise in the future.

2. **Negotiate with yourself as to how many consecutive exercise sessions you must attend to earn this reward.**

   The greater the reward, the more sessions you have to attend.

3. **Write this contract with yourself in your exercise log.**

   See the preceding section for details on what kind of information to put in an exercise log.

4. **When you reach the agreed upon number of sessions, treat yourself — after all, you deserve it!**

Don't get carried away with giving yourself tangible things as rewards for exercising. You don't want to start making yourself think that you are engaging in these healthy behaviors only for the nice little gifts you give yourself. Eventually, you need to develop the view that exercise and a healthy lifestyle are the best gifts you could ever give yourself.

Partnering up

Finding an exercise partner can provide a little extra incentive to work out, even on those days you'd rather not. Research is pretty clear that making a commitment to another person and knowing that person is relying on you is a tremendous incentive to action.

Here's an interesting way to increase the value of the social contract between yourself and your training partner. Every day after exercising together, exchange your workout shoes. If, for some reason, you don't show up for the session tomorrow, your partner won't be able to participate. The same is true if your partner doesn't bring you your shoes. Sure, this tactic adds a little pressure, but it also enhances the commitment between you and your partner to put a high priority on working out.

Remembering that you're doing for yourself, not to yourself

Few things in life more intrinsically rewarding than just having fun. If you want to stick with a long-term exercise or conditioning program, you need to perceive it as something that is generally fun and enjoyable. This entire chapter is devoted to suggestions and topics that can increase motivation and adherence to exercise and make it more fun and enjoyable. As useful as the suggestions in this chapter may be, they pale in comparison to the single most important factor in determining whether you end up hating or loving your exercise involvement: your attitude. Regardless of whether you think exercise is enjoyable or terrible, you'll probably find it to be exactly as you expected.

Chapter 14

Looking Good, Feeling Good: Exercise, Mood, and Mind

In This Chapter

▶ Recognizing the challenges when studying the psychological benefits of exercise

▶ Using meta-analysis to draw conclusions about the effect of exercise on mood

▶ Understanding why exercise makes you feel and think better

Y ou've probably been bombarded with dozens — maybe even hundreds — of messages that give the impression that exercise and physical activity improve your mood and the way the brain works almost as much as it improves the body. But here's the $64,000 question: Are all the messages extolling the psychological, emotional, and cognitive benefits of physical activity really true? And if exercising really does alleviate boredom, anxiety, depression, stress, and tension, and generally make people feel terrific about themselves, why do people have such a hard time sticking with their exercise programs?

The fact is some people do look forward to the positive feelings they get during and after exercising, but many don't. However, the fact that many people fail to experience sufficient emotional pleasures to keep them exercising doesn't necessarily mean exercise is incapable of improving the way people think and feel about themselves. It can — and it often does. But the connection between exercise and emotions is a lot more complicated than most of us ever imagined. In this chapter, we begin to uncover the real story behind how exercise can (but sometimes doesn't) make you feel so great.

Drawing Conclusions from Exercise Research: The Challenges

You probably assume that there's a definitive answer to the frequently asked question, "Does exercise make people feel better?" Alas, that isn't the case. Whether physical exercise is — for any given person, at any given time —

perceived as agony or ecstasy is the result of a virtual Rubik's Cube of unique personal and situational variables constantly interacting with each other in an almost infinite number of complicated ways.

Many of us know, from personal experience, that working ourselves into a huffing, puffing, sweat-dripping lather is an excellent way to reinvigorate not only our weary muscles but also our worried minds. But be careful here — all those wonderful feelings *you* may get from exercise are not necessarily the same things *everyone* would experience from a similar workout. In this section, we outline the difficulties researchers have with answering what seems, on its face anyway, to be a very simple question.

The original Rubik's Cube puzzle that became a world-wide craze in the 1980s has over forty-three quintillion (43,000,000,000,000,000,000 or 4.3×10^{19} in scientific notation) combination possibilities — a number that pales in comparison to the unique biological, psychological, social, and situational variables that impact how people feel when they engage in exercise. For more on those variables, refer to Chapter 12.

Same question but different answers

Explaining why researchers trying to answer the simple question, "Does physical exercise make us feel better?" end up with a completely different set of answers is actually pretty easy. Put simply, exercise isn't exercise isn't exercise, and feeling good isn't feeling good isn't feeling good.

In other words, research studies that have been done so far involve conditions that are unique to each particular study; therefore, the conclusions drawn are unique to those conditions. When you look at different ways to exercise and how many things contribute to whether doing so makes you feel good, it's no great surprise that even good, solid research often comes up with totally different answers.

In addition, the findings are difficult to make any sense of because exercisers, the media, and even some researchers often attempt to draw one-size-fits-all conclusions from a hodgepodge of investigations into a variety of exercise conditions. Every time you exercise in a way that differs from the way a particular study was conducted (and it's almost impossible not to), you're likely going to find it affects you differently.

Following are just a few of the unique conditions that have resulted in more confusion than clarity regarding whether physical activity makes people feel better or not.

How long do I have to keep this up? Optimum duration

Depending on whether you work out strenuously for a couple of seconds, a couple of minutes, or a couple of hours, you'll likely feel quite differently during and after your bout of exercise. The conventional wisdom that you need to keep at it for 20 to 30 minutes to gain the maximum benefits from exercise is largely based on what researchers traditionally thought was required for physiological — not psychological — improvements. And even that is now being called into question.

Exactly how long you have to engage in exercise to experience an improvement in the way you feel or think still remains a mystery. When researchers study exercises that last for varying lengths of time, they'll probably find different answers to the question.

How hard am I pushing myself? Level of intensity

Almost every time you go to the gym, you probably see folks on both ends of the spectrum and everywhere in between. Some people push themselves so hard you expect you're going to have to call 911 at any moment. Others seem to spend most of their exercise time sucking on their water bottles, checking the weight stack, tying and retying their shoes, or preening in front of the full-length mirror. Obviously, the degree to which those being studied work hard and expend energy during physical activity has some bearing on what researchers find — but how much? Trying to draw a single conclusion about exercises of various intensities clouds any research results.

How long have I've been training like this? Current fitness level

A person's general physical condition — basically how fit he or she is — has a lot to do with how exercise makes that person feel. Think about it this way: When you just start a rather strenuous exercise program, you feel pretty wiped out near the end of your session. Worse than that, your muscles are probably screaming at you the next morning as you struggle to get out of bed and shuffle off to the bathroom. However, engaging in this same level of activity after you've improved your fitness level likely produces very different feelings. Not surprisingly, whether research finds that exercise makes people feel good is, to some extent, simply a matter of the fitness level of the exercisers.

How long am I going to feel this way? Duration of benefits

A difference may exist in the length of time you continue to experience any psychological or mental benefits of exercise. Some people find that engaging in certain forms of exercise for precise periods of time under specified conditions results in rather long-lasting (*persistent* or *chronic*) benefits. Others, who exercise in a slightly different way, may only experience positive outcomes that are rather fleeting (*acute benefits*).

Research designed to assess only short-term emotions may very well miss even more important long-term benefits. On the other hand, choosing to measure only long-term effects of exercise may minimize at least temporary positive emotional changes.

What do I do when I'm here? Type of exercise

A question exercise psychologists have been brooding over for quite some time is whether all forms of exercise are truly equal when it comes to making people feel better. The major distinction that has usually been made is between *aerobic* or *cardio* (long duration, moderate intensity) and *anaerobic* (short duration, high intensity) activities. For example, the question may be something like, "Does running for 30 minutes (aerobic activity) make people feel better or worse than lifting weights and doing abdominal crunches for 30 minutes?"

Simply lumping all exercise into one of two incompatible categories — aerobic or anaerobic — is an oversimplification that can lead to a great deal of error and misunderstanding. In reality, any form of physical activity can be placed somewhere along a continuous line drawn between "solely aerobic" and "solely anaerobic," based on the degree to which oxygen transport and consumption is used to fuel the activity. Even a single type of activity — running, for example — may fall on different ends of the scale (running sprints versus running long distances). Confusion and differences over what exactly is aerobic and non-aerobic often results in research studies coming up with different answers to the same question. (Head to Chapter 6 for a detailed discussion on oxygen transport and Chapter 4 for more on aerobic and anaerobic exercise.)

How badly do I need to feel good? Mental state

Physical activity is often touted as something you can do to reduce feelings of stress, anxiety, and depression. What is often neglected in this discussion is that how bad (or good) you feel going into the exercise may have an impact on how the exercise makes you feel. Similarly, how long you've felt this way may also have something to do with whether exercise makes you feel any better.

Thinking that exercise has the exact same effect on people who are only mildly and temporarily depressed as it does on those who have been suffering from deep depression for many years is a mistake. Research results describing changes in mood that result from exercise are likely to be very different, depending on how the subject felt before the exercise and how long he or she has felt that way.

What else can make me feel this good?

Some, but fortunately not all, research findings into the effects physical exercise has on mood, relaxation, and stress are limited because they merely tell us what we already know — that doing something is often better than doing nothing. Getting yourself out of a funk may require nothing more than a bit of a distraction. Playing cards, singing in the choir, meditating, praying, talking with friends, listening to music, or having a nice meal can sometimes make you feel better, too.

Research that concludes that exercise makes you feel better often doesn't consider that it may not be the exercise itself that's making you feel better; doing anything (or many things) is likely to put you in a better mood. Research that doesn't directly compare the results of exercise to other things you could do to improve your emotions is probably telling you more about the benefits of distraction than the benefits of exercise.

Hey, who is that over there? The setting and social component

Physical activity can take place in many social and nonsocial settings. Some people like to kick and punch along with a high-energy Tae Bo video in the privacy of their own homes. Others prefer to sweat it out with a dozen other cyclists in a spinning class or to play an invigorating handball game every lunch hour with a couple of close friends.

Because some exercise is done in private and some involves intense social interaction, it's hard to draw conclusions about the degree to which the specific exercise involved makes us feel. After all, simply interacting with others or taking time by yourself to engage in some much needed contemplation may be all you really need to make yourself feel better. Research that doesn't take into consideration the unique social context within which the exercise is done is often difficult to understand and make sense of.

What am I really looking for? Pinpointing the focus of the research

If you don't know what you're looking for, you probably won't find it, even when it's right in front of your nose. Researchers trying to figure out whether exercise improves the way people think and feel must be able to carefully define, label, and accurately measure the somewhat similar — but also somewhat different — positive thoughts and feelings people attribute to exercise.

For example, a study designed solely to examine stress reduction resulting from physical activity may overlook important changes in exercisers' anxiety levels. Similarly, a study that specifically focuses on changes in feelings of depression may not see the benefits of exercise on anxiety. It doesn't mean these positive effects didn't happen; it just means the researchers weren't looking for them.

Researchers can measure only those things they choose to measure, and their research can find only what it sets out to find. If you go beyond the parameters of the study or if you are fuzzy on what mood states the research was designed to test (you assume that a reduction in depression, for example, is the same as a reduction in stress), you end up drawing the wrong conclusions as to whether exercise alters important mood states and what states it affects.

Crunching the Numbers: Meta-analysis

Despite the fact that the terms *exercise* and *physical activity* mean different things to all of us, exercise psychologists have begun using powerful research tools to help them figure out, at least to some extent, the conditions under which exercise can improve mental outlook. One very useful statistical technique is called a *meta-analysis.* Although you may find some of the number crunching involved in meta-analysis a bit confusing, the basic idea behind what it does makes a lot of sense and is actually pretty easy to understand.

Meta-analysis is just a fancy term that means to do an analysis of analyses. Basically, it's a clever statistical trick that lets researchers combine and summarize a whole bunch of separate research findings to reveal overall trends that may not be apparent in the individual studies alone.

Looking at how meta-analysis works

To conduct a meta-analysis, researchers take similar studies, run some calculations on the combined data, and then evaluate the results to see whether any meaningful trends emerge. Here's a simple example:

Researcher A conducts a study to analyze the effect of aerobic (long duration, moderate intensity) exercise on feelings of stress. She has study participants run on a treadmill for 15 minutes in a way that maintains about 75 percent of their maximum heart rate throughout the activity. Before and after their stint on the treadmill, the participants each complete a short questionnaire that asks about how much stress they are experiencing at that moment.

A year or so later, Researcher B, who is interested in roughly the same general question as Researcher A, designs a similar study in a slightly different way. He decides to have his participants ride on an exercise bike during a structured, instructor-led, hour-long spinning class and gives them a somewhat different — but comparable — questionnaire assessing their stress levels before and after the class.

The two studies are obviously a little bit different (they used different types of exercise for different duration and measured stress in different ways), but they also have a lot in common. Both studies set out to look at the effect of an aerobic activity on exercisers' perceived stress levels. Both researchers had participants rate the level of stress they were experiencing before and after they exercised so that changes in stress levels that occurred during the activity could be examined.

If you were to analyze each of these studies separately, it's possible that you would find that neither the runners nor the spinners felt much difference in the reduction of stress. If that happened, could you definitively say that aerobic activity doesn't relieve stress? Well, not so fast. One reason you may not see a difference is that perhaps neither study, by itself, had enough subjects to actually yield a statistically significant effect, even if one truly existed (researchers often refer to this as a lack of statistical *power*).

However, by combining the findings of these somewhat different studies (and as many more like them as you could find) and conducting one gigantic analysis (a meta-analysis) of all the separate analyses, you may discover something very different.

Very often, combining the results of many studies through meta-analysis allows researchers to see important overall trends that wouldn't be apparent simply by examining each study individually. Of course, merely conducting a meta-analysis doesn't guarantee that the true relationship between exercise and emotions will magically be uncovered.

Uncovering limitations of meta-analysis

As powerful and useful as a meta-analysis can be in helping researchers better understand how exercising changes the way people think, it's certainly not perfect. Like all tools, a meta-analysis itself has a number of limitations, all stemming from the fact it's only as good as the people who use it:

- **Someone has to decide whether different studies are truly comparable.** Do the studies have enough in common to justify clumping them all together? In the previous example, what if Researcher A had adolescent males with mild cognitive disabilities run on the treadmill and Researcher B tested a spinning class that only included females over the age of 50? Would you consider these two groups of exercisers similar enough to be combined into one single super analysis? Some people may say yes; others may disagree. This determination is often a judgment call with no right or wrong answer.

Researchers must ensure that all relevant studies, and not just those that support their hyphotheses, are included. Yet the subjective nature of deciding what to include and what to leave out can lead to a big problem: Some researchers get so focused on proving their hypotheses that they exclude, either intentionally or unintentionally, studies that are not

likely to support their positions. See the next section for more information about how the studies reviewed can influence the reliability of the
conclusions.

✔ **Researchers have to know that a study exists before they can include
it in the meta-analysis.** This point may sound like a no-brainer —
something exists or it doesn't, right? — but it also turns out to be a bit
more complicated than it sounds, because of the things that go into
decisions regarding whether to report findings. If researchers include
only reported or published studies, they may unintentionally miss
relevant studies that didn't get published for one reason or another.

The following sections take a closer look at these problems.

Picking only the tastiest "cherries"

Sometimes, researchers are so intent upon demonstrating that their ideas are
correct that they — either consciously or subconsciously — disregard studies where the results don't seem to support their pre-existing position. (This
tendency is sometimes called *cherry-picking* the research you want to include
in your meta-analysis.) Although all studies differ in some way or another, the
person conducting the meta-analysis has to determine whether these differences are still close enough to justify clustering them together or not.

The decision to include or eliminate a study from a meta-analysis may not as
quite as straightforward as you think; it can even sometimes be influenced by
what you believe the answer "should" be. Consider the earlier example and
start with the assumption that, in your heart of hearts, you truly believe that
aerobic exercise relieves stress. When you come across the first study (the
15-minute run on the treadmill), you find, much to your disappointment, that
the running did *not* reduce the subjects' perceived stress levels. They were
pretty much as high after running as they were before the subjects got on the
treadmill.

When you read the study more carefully, however, you notice that the instant
the runners stepped off the treadmill, the researchers also did a blood draw
to measure the runners' lactate levels. Because of your personal biases (you
expected to see a reduction in stress after exercising), you may conclude that
the mere act of jamming a needle into a person's arm is stress-provoking in
itself, thereby contaminating the results of the study. If you decide that the
results of the study are, therefore, not believable, you may feel justified in
eliminating it from your meta-analysis. But what if the study — despite the
taking of a blood sample — *did* find a reduction in stress levels after exercise,
as you thought it would? Would you now be more inclined to include it in
your meta-analysis? If you would exclude it in the first instance but include it
in the second, you're cherry-picking the data.

When determining how sound the conclusions of a meta-analysis are, check to see whether the researchers describe in detail exactly how and why studies were either included or excluded from the analysis.

The file-drawer effect

Occasionally, findings that conflict with the expected results are excluded from a meta-analysis, not because they are intentionally eliminated, but because the person doing the meta-analysis never knew the research existed in the first place.

Researchers typically conduct a lot of studies; sometimes their studies find a positive result, sometimes they don't. If the study comes out supporting the researcher's initial hypothesis, then it's likely he or she will write it up and send the manuscript off to be considered for publication in a professional journal. If, on the other hand, the results don't come out the way the researcher thought they should, there is a tendency on the part of researchers to assume they've messed something up and simply throw these negative findings in their "file drawer" and move on to more interesting (and potentially rewarding) scientific investigations. If this happens, all the findings that failed to find what researchers expected never end up seeing the light of day and, therefore, can't be included in a subsequent meta-analysis on the subject.

When you are trying to figure out whether the conclusions of a meta-analysis are believable, be sure to examine the extent to which the researchers attempted to locate studies that might have been overlooked due to either the file-drawer effect or a publishing bias (explained in the next section).

Publishing bias: Who's interested in reading about nothing?

In a slight twist on the file-drawer effect, where the researcher himself or herself decides that the unanticipated results of a study are probably not worth sharing, professional journal editors and their editorial boards typically act as gatekeepers, deciding whether a particular research finding makes enough of a contribution to science to warrant publication.

Even if a researcher finds the courage to pull a "non-significant findings" manuscript out of the file drawer, dust it off, and send it to a peer-reviewed journal, it may not be accepted for publication. When research isn't published, for any reason, it becomes virtually impossible to include in a meta-analysis that may shed light on the relationship among exercise, feelings, and thoughts.

As protectors of the accepted scientific knowledge in their field, these people and groups are often reluctant to publish articles that have arrived at the conclusion that basically say, "Nothing really happened in this study."

How interested, for example, would a professional journal catering to exercise and fitness specialists be in publishing an article that concludes that physical activity has no positive effect on the way people feel? Such a finding not only runs counter to what many people (especially the readers of this particular journal) intuitively believe, but it also is at odds with the results of previously published research. Furthermore, it's difficult to actually prove a negative — that is, to prove that something did *not* happen.

Almost any type of mistake or contamination at any point in the study protocol (like drawing blood from the subjects in the earlier example) can cause the expected outcome to be lost.

What We Think We Know about Exercise, Emotion, and Cognition

Despite the mixed research findings and the built-in limitations of meta-analysis, many exercisers — and most researchers — are thoroughly convinced that strenuous physical activity can produce some pretty impressive improvements in the way people think and feel. Abundant anecdotal and increasing scientific evidence support the notion that regular exercise has the potential to enhance body image and self-esteem while reducing perceived stress, anxiety, and depression.

If you go into almost any gym or fitness center in the country and ask those who are there working up a real sweat, chances are they'll emphatically tell you how great they feel when they exercise. And there's certainly no reason to doubt them. After all, this improvement in their mood is probably the thing that, in large part, keeps bringing them back to work out again and again.

In addition, recent evidence also suggests that physical activity can, in addition to making exercisers feel good about themselves, potentially improve cognitive functioning and possibly even postpone or prevent debilitating effects of aging, such as Alzheimer's disease and dementia.

After listening to regular exercisers (or yourself, if you happen to be one of these advocates of exercise), it's easy to come away with the impression that exercise obviously, and almost always, has a positive effect on mood. Is that impression accurate? Read on to find out what the scientific research has to say.

Measuring mood with the Profile of Mood States

Accurately measuring a person's mood is a pretty complicated process. Because emotions only exist somewhere deep inside a person's head, there's no way you can ever really know what someone else is feeling at a particular time unless you ask. When measuring moods that are associated with exercise, researchers generally have little choice but to rely on self-reports; they simply ask people to tell them about the feelings they experience before, during, and after exercising.

Because researchers have to rely so heavily on exercisers' descriptions of how exercise makes them feel, they must have accurate ways of collecting this information. That's why researchers like to use psychological measures that do following:

✔ Assess a broad range of feelings

✔ Are generally considered by experts from a variety of disciplines to be reliable and valid indicators of the way people truly feel

✔ Have been used as a way of measuring mood in a large number of clinical and exercise-related settings

Although several paper-and-pencil questionnaires have been used over the years to assess mood, the one psychological tool that has been used far more than all the others combined is the *Profile of Mood States,* commonly referred to simply as the *POMS.* Literally thousands of studies have used the POMS as an accurate measure of mood.

The six mood states the POMS assesses

The original POMS consists of 65 adjectives such as "relaxed," "nervous," "annoyed," and "energetic," rated on a five-point scale (0 = Not at all; 4 = Extremely) indicating the extent to which a person thinks this particular descriptor accurately describes the way he or she feels. The POMS items are designed to assess six general mood states; these subscales are

(1) Tension

(2) Depression

(3) Anger

(4) Vigor

(5) Fatigue

(6) Confusion

Notice that, of all the emotional states measured, only one is positive: vigor. Knowing this is a key component of evaluating what the POMS results tell you.

Showing the signs of a very cool customer: The Iceberg profile

After the subject completes the POMS survey, the researcher plots the scores from each of the six POMS subscales, and the result gives a picture, both literally and figuratively, of the subject's mood. If you start to see something that looks a bit like an iceberg (see Figure 14-1), the person is probably in a positive and resourceful emotional state. Not surprisingly, this "high energy" iceberg profile has frequently been shown to be associated with high levels of athletic performance.

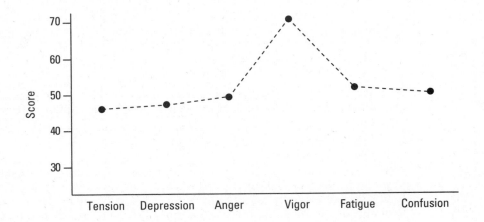

Figure 14-1:
An iceberg indicates a positive emotional state (top); a flatter line indicates a less positive mood (bottom).

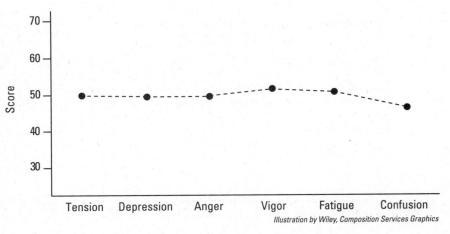

Illustration by Wiley, Composition Services Graphics

The reason the "productive" profile resembles an iceberg is pretty simple. To produce such a graph, the scores measuring a person's perceived levels of tension, depression, anger, fatigue, and confusion (all negative feelings) are relatively low, but the score for vigor, the one and only positive feeling measured by the POMS, is high. The result is a huge spike right there in the middle, causing the profile to take on the appearance of a giant iceberg.

Conversely, if you plot out someone's relatively high scores on the five negative scales and a low score in vigor, you produce a profile picture that appears basically flat. You can easily see why this "melted iceberg" isn't a very good sign. It indicates the person feels rather high levels of tension, depression, anger, fatigue, and confusion and, at the same time, very little vigor or energy.

Uncovering the connection among stress, anxiety, and exercise

Although exercise psychologists don't view stress and anxiety as being completely synonymous, they do think that these mental states are related in several ways and that both can be reduced with regular physical activity.

Differentiating between stress and anxiety

You probably think of stress and anxiety as being more or less the same thing. Exercise psychologists, however, don't. Here are the key, albeit subtle, differences:

- **Stress:** *Psychological stress* is usually defined as that horrible feeling you get when demands are placed on you that you really don't think you can meet. Not only that, you also believe serious negative consequences will occur if, for whatever reason, you fail to meet any of these demands. Basically, when you feel stressed, you feel overwhelmed. In modern society, much of the stress people feel is caused by thinking that not enough time exists in the day to get everything done and that bad things are going to happen if they let anything fall through the cracks.

- **Anxiety:** *Anxiety* differs from stress in that it is merely a negative feeling of worry, apprehension, or nervousness for any reason. You can be anxious about flying, anxious about a burned dinner, anxious about upcoming test results, anxious about giving a presentation at work; you can even be anxious about nothing in particular (what psychologists sometimes call *generalized anxiety disorder*).

Stress and anxiety sometimes feed off one another. If, for example, you're feeling overwhelmed and unable to meet your obligations, you're likely to start feeling worried or nervous — in other words, anxious. On the other hand, the time you waste worrying about things may further reduce your ability to meet your obligations, and then you become stressed. The result is often a vicious circle of stress and anxiety, anxiety and stress.

Lessening anxiety with exercise

The reason recognizing the subtle difference between stress and anxiety is important is because exercise is likely to be a bit more effective in getting you to stop worrying than it is in helping you meet the excessive demands that are placed on you.

Although many factors can alter the results, after examining all the available research findings so far, researchers can safely say a few things about the impact exercise can have on anxiety levels:

- Exercise is associated with small to moderate reductions in feelings of anxiety.

- Exercise is at least as effective — and sometimes more effective — in reducing anxiety than other commonly used treatments for reducing anxiety, such as talk therapy and mild anti-anxiety medications.

- Exercise tends to have beneficial effects on both chronic (long-term) as well as acute (short-term) feelings of anxiety.

- All forms of exercise appear to have the potential to reduce anxiety, but the largest effects are usually seen with aerobic (moderate intensity) exercise that lasts at least 30 minutes.

- Reductions in anxiety levels appear to be possible even with intensities between 30 percent to 70 percent of maximal heart rate.

- Your mood will generally return to its original state within 24 hours or so after exercising, meaning that it's important to exercise on a regular basis to keep these feelings of anxiety at bay.

Exercising to relieve depression

Although everyone occasionally feels a little down in the dumps because of a personal or professional setback, if these feelings last for an extended time, the real culprit may be at least a mild form of depression.

As is the case with anxiety, certain types of exercise show great promise in at least minimizing the negative effects of depression. Several studies have shown that a well-designed exercise program is just as effective as traditional psychotherapy or other behavioral interventions in reducing depression. Although that fact may not sound like such a big deal, it is, for these reasons:

- Exercise is generally much less expensive than traditional psychotherapy.

- Exercise doesn't carry with it the social stigma that unfortunately is often part of being in psychotherapy.

How depressing!

Feeling depressed? It wouldn't be all that unusual if you were. Depression impacts the lives of millions of people a year. Consider these stats:

✔ The Centers for Disease Control and Prevention (CDC) has estimated that about 10 percent of adults in the United States are suffering from symptoms of depression at any given time.

✔ Somewhere between 15 percent and 20 percent of Americans will suffer from a major depressive episode at some point in their lives.

✔ Depression is a world-wide problem with probably somewhere in the neighborhood of 300 million people around the world suffering from some level of depression.

Because of the sheer numbers of people affected — and the severe disruption it creates in virtually every aspect of their lives — depression is becoming the single major cause of disability in the United States. Needless to say, exercise scientists are very eager to discover the conditions under which physical activity may play a role in at least reducing the impact of this terrible disorder.

✔ When compared to other behavioral interventions used to treat depression, exercise is the only way to gain all the other benefits generally associated with increased physical activity, such as improved cardiovascular health, sustainable weight loss, strengthening of muscles and bones, and enhanced coordination and flexibility.

Not surprisingly, many of the same characteristics that seem to make exercise useful in reducing anxiety also contribute to its effectiveness as a potential treatment of depression. For example, preliminary findings suggest that the longer the physical activity program lasts, the greater its effect on depression.

Understanding Why Activity Helps You Beat the Blahs

Nearly every day, rigorous scientific investigations uncover more and more evidence supporting the idea that a regular exercise program can significantly improve the way people think and feel about themselves. As exhilarating as it is to be on the cutting edge of these breakthroughs, researchers still have a lot of work to do. For one thing, we don't really understand why or how physical activity alters emotions. Oh sure, there are plenty of guesses, assumptions, hypotheses, and ideas about how exercise is changing the way we think and feel, but scientists still have a lot more questions than answers. Much of the thinking to this point has centered on physical activity's ability to change the body's chemistry.

Your body's own little (legal) drug lab

When you exercise, your body works really hard, but that's not the half of it. Sure, all the jumping and pushing and shoving and bouncing and huffing and puffing and flushing and sweating are visible on the outside, but what you can't see are all the tiny chemical changes deep inside your body that even short bouts of moderate exercise ignite.

Currently, scientists are really interested in four particular chemicals that the body manufactures: serotonin, norepinephrine, dopamine, and endorphins. They think these chemicals may be responsible for at least some of that "feel good" feeling people often get from exercise.

Serotonin: The brain's feel-good chemical

Serotonin is a *neurotransmitter* (a chemical that is sent from one nerve cell to another) that has been shown to be associated with pleasant sensations. Basically, the more serotonin you have in your brain, the better you feel. In fact, most antidepressant drugs are intended to help cells send as much serotonin as possible on to neighboring cells.

Although most of the research done so far has been conducted on animals, these preliminary studies make researchers think that getting involved in physical exercise jacks up serotonin levels in the brain and makes the exerciser feel better. Exercise seems to be doing a couple of things that may raise serotonin levels and decrease depression and anxiety:

✔ It may directly simulate the release of serotonin from the cells.

✔ It may indirectly raise the level of another chemical — tryptophan — which is a building block in the manufacture of serotonin.

Norepinephrine: The stress hormone that relieves stress

Chances are you think of epinephrine (also called *adrenaline*) as the hormone that your body releases in response to stress. You've probably heard the stories about people performing superhuman feats (like mothers picking up automobiles to rescue their trapped infants, for example) because they have a large amount of epinephrine surging through their veins.

When you're engaged in super-intense exercise — or any other stressful event — your supply of epinephrine can be depleted, and you can feel exhausted, mentally fatigued, and depressed. At moderate levels of exercise, however, epinephrine levels are lowered, and you're likely to experience decreased heart rate, body temperature, and blood pressure, and a calming sensation.

Don't get confused when you hear people using the labels *adrenaline* and *epinephrine* (or *noradrenalin* and *norepinephrine*) to describe what appears to be the same fight-or-flight hormone: They are *exactly* the same! The only difference is where the two words originated: *Adrenaline* comes from Latin origins, and *epinephrine* is based on Greek.

In Latin, *renal* means kidney, and the Greek term for kidney is *nephron*. So both *adrenaline* and *epinephrine* are really just pointing you — one in Latin, one in Greek — to where most of this hormone is produced: in the adrenal medulla (which simply means in the middle of the adrenal gland) that sits on top of your kidneys. The Latin-based term *adrenaline* is used more often around the world, but American health professionals and scientists typically prefer the Greek-based *epinephrine*.

Dopamine: Rewards for the brain

Like serotonin and norepinephrine, *dopamine* is a neurotransmitter that has been shown to both increase during exercise and have a considerable impact on mood, pleasure, and motivation.

Dopamine is the chemical that the brain uses to reward itself. When things are going well — when you think you're going to get something you like — your brain gives you a big shot of dopamine as a reward. This reward makes you more likely to remember what you did and repeat it. Just as with serotonin and norepinephrine, exercise causes more transmission of dopamine; with more dopamine available, mood and pleasure increase.

Endorphins: Runner's high — or urban legend?

Of all the exercise chemicals that scientists think may cause people's good feelings, endorphins are the ones that most exercisers think of first when they're asked to explain why exercise gives them a euphoric sense of well-being. *Endorphins* are often credited with causing the well-known phenomena of the "runner's high," the feeling of ecstasy and well-being that sometimes accompanies long, rhythmic forms of exercise.

Scientists originally thought that endorphins were responsible because they are essentially the body's own private stash of opium. In fact, the term *endorphin* itself is just a combination of two words that basically tell us the whole story: _endogenous_ (meaning a substance that originates within the body) and _morphine_ (a powerful narcotic pain reliever). Because the levels of this self-made morphine in the blood were known to rise dramatically during exercise, it seemed perfectly logical to assume that they were responsible for any feelings of exhilaration and elation that unexpectedly washed over exercisers in the midst of strenuous physical activity.

As scientists looked closer, however, they started questioning whether these naturally produced drugs were really capable of causing such euphoria during exercise. Here were the main concerns:

✔ **Doubts existed about whether all the endorphins seen floating in the blood could even enter the brain at all.** Endorphins appeared to be too big to squeeze through the _blood-brain barrier,_ which acts as a sort of border patrol guard keeping unwanted junk that's swimming around in the blood from entering the brain and causing big trouble.

✔ **Some research has found that even if endorphins were somehow able to get across the blood-brain barrier, it might not matter anyway.** These studies discovered that exercise can still produce joyous, euphoric feelings even after a chemical block is administered that makes it absolutely impossible for any endorphins to get into the brain.

These observations caused many experts to give up on the idea that endorphins were responsible for the runner's high. Recently, however, researchers in Germany, using special tracking techniques and medical imaging (PET scans), reported spotting endorphins in the brain. Scientists aren't sure how the endorphins got there, but it's clear that we're going to have to rethink the role of endorphins and the "runner's high" — again.

Investigating other reasons exercise may make you feel good

Until researchers locate some hard evidence that the body's naturally produced "drugs" are responsible for the great feelings some people get when they exercise, they continue to explore what else may be happening during workouts that could make exercisers feel so good. Although it's impossible to be 100 percent sure of anything yet, following are some strong possibilities.

Increasing body temperature

One possible reason people sometimes feel so good while exercising — aside from the drugs the body produces — is that even moderate physical activity can increase the temperature inside the body and cause changes in blood pressure. The positive feeling associated with exercise may simply be a function of increases in temperature, blood flow, and blood pressure. In short, during exercise your body generates its own warm, cuddly feeling!

Benefiting psychologically by getting into the flow

One commonly held notion regarding why people sometimes feel so great when they engage in physical activities has to do with their ability to get "into the flow" or "in the zone."

According to this theory, when you're engaged in a task that is a perfect match for your skills and abilities — it's neither too easy nor too hard — and you're performing for intrinsically motivating reasons (and not to collect external recognition or rewards), you are sometimes able to intensely focus on the activity to the point that you experience an altered state of consciousness, in which time slows down and you feel totally competent and completely in touch with yourself and your surroundings.

Although this explanation doesn't describe what's happening at the cellular level, it still provides a pretty good practical idea of how your thoughts can influence the way you think and feel.

Enhancing body image and self-esteem

It's easy to get all caught up in the physiological, chemical, cellular, and molecular reasons exercise makes people feel good, but don't discount the many other indirect, but equally important, reasons why physical activity can make you feel more positive about yourself and the world around you. Simply put, understanding what exercise does *to* you may not be as important as understanding what it does *for* you, or how it can indirectly improve virtually every aspect of your life:

✔ **You feel better when you feel healthy.** There is absolutely no doubt that regular physical activity can get you — and keep you — fit and healthy. When it does, it's naturally going to make you feel better about yourself and your life in general. Remember that old saying, "The best part of health is a good disposition."

✔ **You feel better when you look better.** Exercise — if done regularly and for an extended period of time — can help you look better. And when you look better, you've taken a giant step toward feeling better. When you convert a few pounds of flab to a couple pounds of solid muscle and find yourself fitting back into all those great looking clothes that have been hiding in the back of your closet for years, you'll start to feel really good about yourself. As your self-confidence improves, you'll be willing to engage in more fun activities, which will make you feel even better about yourself.

✔ **You feel better when you are successful.** Starting out on and sticking with an exercise program can result in a tremendous sense of personal achievement; and it's something you can — and should — be very proud of. Knowing that you've not only set but also followed through on a life-changing goal can boost your self-esteem and confidence. Of course, it doesn't hurt your feelings of self-worth and accomplishment when others recognize your successes, too.

✔ **You feel better when you do more.** By becoming physically active, you are increasing the likelihood you'll become even *more* physically active. It's a positive feedback loop.

By engaging in regular physical activity, you can improve your health, improve your appearance, and have the self-satisfaction of accomplishing important, life-altering goals. All these personal improvements will motivate you to enjoy life even more.

Growing Bigger, Stronger, Faster . . . and Smarter, Too!

A couple thousand years have come and gone since the ancient Greeks and Romans popularized the idea of developing a strong mind *in* a strong body, yet only in the last couple of decades have researchers started to realize that developing a strong body *might actually lead to* developing a strong mind. Relying heavily on encouraging preliminary findings from human research and extrapolations from hundreds of animal studies, scientists are starting to see that the benefits of exercise may not be limited to improvements in health and emotions. Working up a good sweat may even play an important role in the way kids and adults learn, remember, and process information. Exercise may actually help us get smarter and stay smarter.

Building up the hippocampus

Although scientists once thought that humans are born with a finite number of brain cells and that dead brain cells could never be replaced, recent research has shown that, at least in mice, voluntary exercise (on a running wheel) is associated with the production and survival of brand new cells in the *hippocampus,* the part of the brain responsible for memory.

This finding is important not only because it suggests that exercise can cause the brain to make and nurture new brain cells that are involved in memory, but also because it has huge potential implications for humans. Why? Because the hippocampus, where these new cells are growing, is the part of the brain that shows the very first signs of damage in people who suffer from Alzheimer's disease.

No one is precisely sure why simply giving mice the opportunity to exercise helps them make new brain cells and strengthen the connections between existing nerves in the brain. Something as simple as increasing blood circulation to the brain may be at least part of the answer. Or perhaps exercise increases the supply of serotonin, norepinephrine, or dopamine in the brain (refer to the earlier section "Your body's own little (legal) drug lab" for information on these chemicals). Considerable evidence exists that exercise can increase the levels of *brain-derived neurotrophic factor (BDNF)*, a protein that strengthens neural connections while protecting and growing new nerve cells.

The word *hippocampus* mean "seahorse." This little structure in the brain primarily responsible for converting your daily short-term memories into more permanent long-term memories got its strange name because Julius Caesar Aranzi, a famous surgeon and anatomist at the University of Bologna in the mid-1500s said that's basically what the thing looked like to him — a seahorse.

Leaving no child (on his or her) behind

In January of 2002, President George W. Bush signed into law a piece of legislation commonly known as "No Child Left Behind." The idea was to get schoolchildren to learn more by increasing standardized testing and holding both teachers and schools more accountable for making sure students were learning at a higher level.

What's interesting is that preliminary research suggests that, if we were truly concerned about helping kids learn more, one of the first things we'd do is set aside plenty of time for them to engage in physical exercise, both in and out of school. Ironically, despite consistent findings indicating that kids who spend more time exercising tend to do better in school, school districts are under more and more pressure to cut physical education classes in an attempt to save money.

Staying fit as a fiddle and sharp as a tack

As we age (and all of us will, if we're lucky), our brains tend to slow down. Cells die off and are rarely, if ever, replaced. People for whom this wearing out of the brain occurs faster than normal begin to lose the ability to remember, pay attention, make decisions, and even speak. The terrible nature of this condition is readily apparent by simply looking at its name: *dementia,* which in Latin literally means "without a mind."

Although several forms of dementia exist, the most prevalent — and the one you've no doubt heard of — is Alzheimer's disease. In this devastating disease, abnormal clumping (amyloid plaques) and tangling (neurofibrillary tangles) of nerve cells in the hippocampus of the brain basically stop a person's ability to think properly or remember things. The overall connections between the brain cells also break down, causing the cells to stop sending messages properly and even to die. If exercise can replace some of these dead or damaged cells in the hippocampus with new and functioning cells, we may be able to stop, or at least slow, the terrible effects of Alzheimer's disease.

It's estimated that about 5 million Americans — and their families — currently endure the ravages of this incapacitating disorder. Worse yet, some epidemiologists predict that, because of the large number of aging baby boomers (people born between 1946 and 1964) and increases in life expectancy, a shocking 13.8 million Americans and 1 in 85 people worldwide will have Alzheimer's by the year 2050, unless researchers can somehow find a way to reduce the number of people affected or postpone the onset of the disease.

Research, including that currently being conducted at the National Institute on Aging, is very encouraging thus far. In the few studies that have looked at the effects of exercise on the cognitive functioning of the elderly, most have found that increased physical activity tends to improve or at least maintain brain functioning. In other words, if people — especially those who are older — start exercising (and making additional brain cells in the hippocampus), they may be able to at least minimize the negative impact of aging on cognition.

If simply getting people to exercise early, often, and on into their later years can reduce the debilitating effects of various types of dementia, kinesiologists will have played a major role in one of the greatest improvements on peoples' lives in the history of the world.

Chapter 15

Too Much of a Good Thing

Although we all know that regular exercise can have many physical and psychological benefits, when taken to extremes or when practiced in combination with certain environmental conditions or personal risk factors, even this typically healthy, positive behavior can be bad for you. While the admonition to exercise regularly is usually very solid advice, a few rare, but important, exceptions to this general rule exist.

In this chapter, we explore how something that is usually one of the best things you can ever do *for* yourself may, in some extraordinary cases, degenerate into one of the worst things you can do *to* yourself.

Good Exercise Gone Bad: Understanding the Problem of Excessive Exercise

For a variety of reasons that are not fully understood, some people (usually estimated to be about 3 percent of the general population) develop the unhealthy craving to take their exercise to such extremes that it begins to do them more harm than good. This intense urge to exercise can have serious physical, psychological, and social implications. Increased likelihood of physical injury, psychological burnout, and the neglect of family, friends, and work are just a few of the negative consequences that are often associated with excessive amounts of exercise.

When is an addiction not an addiction?

For years, mental health professionals have engaged in an ongoing debate as to whether or not excessive behaviors, such as the tendency to over-exercise, should be referred to as *addictions, obsessions, compulsions*, or something less medical-based entirely. The latest skirmish in this ongoing battle erupted during the recent revision of the American Psychiatric Association's *Diagnostic and Statistical Manual of Mental Disorders (DSM-5),* which is used to diagnose and classify mental disorders.

In the final edition, the section "Addictions and Related Disorders" is generally limited to compulsions toward drug-seeking behaviors.

The new DSM-5, however, includes a broad category of disorders called "behavioral addictions," which, oddly enough, lists only one behavioral disorder — gambling addiction.

The good news is that the addition of this new category will, in itself, probably increase attention and research on other potential behavioral disorders — including the need to exercise too much. The bad news is that not being specifically mentioned in the DSM-5 will make insurance reimbursements virtually impossible for health professionals wishing to work with clients who exhibit symptoms associated with excessive exercise.

This extreme need to exercise goes by many names. Although often informally referred to as *exercise addiction,* most researchers and health professionals prefer less emotionally charged terms such as *exercise dependence, exercise compulsion, dysfunctional exercise, excessive exercise,* or (heaven forbid) *hypergymnasia.* Regardless of what it's called, pinpointing exactly where good and healthy physical activity becomes harmful is almost impossible.

 Very few things are better for the human body than physical activity and exercise. Most people simply don't get enough of either. Because of the emphasis that is frequently placed on exercising, people sometimes forget that, like pretty much everything in life, there is a point at which things that are generally good for you can sometimes do you great harm. So that you can take full advantage of all the good things a physically active lifestyle can offer without having to deal with the problems that sometimes come when you take this good thing too far, carefully think about where, when, why, and how much you exercise.

Crossing the line: How much is too much?

The thing that makes any discussion of excessive exercise so tricky is that researchers and health professionals don't really know — or at least can't seem to agree on — what they're even talking about. If pressed, most people distinguish between beneficial and excessive exercise with something like, "If I schedule my exercise so it fits into my life, that's a positive and healthy thing, but as soon as I start scheduling my life so it fits into my exercise, I may be on the verge of taking this whole thing a bit too far."

Exercise specialists have agreed that the best way to identify and assist people who may be exercising too much for their own good is to consider the reasons people exercise and the positive or negative impact it has on their lives rather than on how much exercising they actually do.

Identifying symptoms of excessive exercise

Researchers still, for better or worse, tend to think about behavioral excesses in much the same way they think about addictions. In fact, the most common ways to measure whether someone has gone overboard with exercising is to see responses to a series of questions that ask the person to describe what he does and why he does it — the same kind of survey used to figure out whether someone is addicted to cocaine, heroin, alcohol, or some other chemical substance.

Table 15-1 lists the seven symptoms that are usually examined when determining whether a person is engaging in healthy or unhealthy exercise and the kind of questions asked to determine whether someone has a problem or not. It's pretty easy to see that they were intentionally developed to mimic the way addictions are assessed.

Table 15-1	**Symptoms of Excessive Exercise**	
Symptom	*This Question*	*Is an Indication of This*
Tolerance	Do you need to keep doing more and more exercise just to get the same good feelings?	Whether you are building up an exercise tolerance.
Withdrawal	Do you feel anxious, irritable, or have trouble sleeping when you aren't able to get your regular amount of exercise?	Whether you're experiencing withdrawal symptoms when you can't get your daily exercise "fix."
Intension	Do your workouts last longer or involve a great deal more effort than you had originally intended?	Whether you are unable to limit yourself to your intensions regarding exercise.

(continued)

Table 15-1 *(continued)*

Symptom	This Question	Is an Indication of This
Control	Do you wish you could — but don't seem to be able to — cut down on the amount of exercising you do?	Whether you lack the ability to control your exercise.
Time commitment	Do you spend a lot of time exercising or planning what you'll do when you exercise?	Whether you make large time commitments to exercise.
Interference	Have you had to reduce or eliminate other activities — social, recreational, personal, or job-related — to accommodate your workout schedule?	Whether you are reducing involvement in other activities in order to exercise.
Continuance	Do you continue exercising even when you know it's probably not good for you — like when you have a physical injury or illness?	Whether you continue to exercise even when you know it may result in negative consequences.

Answering yes to one or even a couple of these questions doesn't necessarily mean you have a serious exercise problem. However, the more these experiences describe you and the way you exercise, the more you may want to begin thinking about whether you're placing just a bit too much emphasis on your training at the expense of other important things in your life.

Examining why people exercise too much

Most people do or don't do things as in order to either increase pleasure or to decrease pain. Deciding whether to exercise or not is no exception. Let's face it: Most people probably wouldn't exercise at all if they didn't really think that all that hard work would either pay off in some positive way or help them avoid or delay something negative. This desire to maximize pleasure and minimize pain may drive how often people exercise and whether they end up developing an over-attachment to exercise.

Getting hooked on the feeling

Ask people why they engage in intense, regular exercise, and many tell you, in almost spiritual terms, about the fantastic feelings of well-being they get when they physically push themselves to their limits.

As we explain in Chapter 14, intense exercise can produce chemical changes in the body that are often associated with feelings of euphoria. In fact, a couple of the chemicals that the body releases during strenuous exercise (serotonin and norepinephrine) are so closely linked to "feeling groovy" that they serve as the basis of most commercial antidepressant medications. Two other chemical byproducts of exercise — dopamine and endorphins — are essentially the body's own way of treating itself to a reward.

With all those pleasure-producing drugs being pumped into the brain during long and hard workouts, it's no surprise that some people continually seek out the amazing sensations they can get (legally) only through intense exercise. Anyone who develops a craving for extreme exercise just for the sheer pleasure it brings is said to have a *primary,* or an immediate, attraction or dependence to exercise. But there are other, secondary, reasons some people can't seem to get enough of this good thing called exercise, which we explain in the next section.

No pain, no gain: Believing that the ends justify the means

Some exercise extremists don't find exercise for its own sake enjoyable in the least. Instead, they are far more interested in what they think their strenuous workouts will do for them in the long run. To them, exercise is little more than a means to an end, the price that must be paid to get something they want in the future.

Athletes, for example, may routinely drive themselves to the point of exhaustion in the hopes of seeing just the slightest improvement in their performance. Those who are particularly vulnerable to this type of motivation are

- Those who participate in events that are judged, in part, on the way the performer looks: gymnasts, figure skaters, dancers, competitive cheerers, and so on.
- Those who participate in sports with severe weight restrictions or classifications: wrestlers, boxers, mixed martial artists, and equestrians, for example.

These people use excessive exercise as a way to control their weight, alter their appearance, and thus improve their performance.

Even non-athletes often push themselves through countless hours of demanding physical workouts just to drop a few pounds, develop those "iron abs" or "buns of steel" that they can then show off during their next trip to the beach. Their motivation is to work hard in the gym now in order to reap the benefits sometime in the future.

Meaningless and misleading words: "Anorexia athletica" and "sports anorexia"

Some people use invented and highfalu-tin' terms like *anorexia athletica* and *sports anorexia* to make a connection between excessive exercise and a specific disordered eating pattern (anorexia). The problem with this combined term is that it doesn't make sense: *Anorexia is a* combination of *an* (a lack of) *orexia* (appetite), and so it literally means a lack of appetite. Although the exercise and the lack of appetite may be happening at the same time and be designed to accomplish the same goal (weight loss), having a lack of appetite really has nothing to do with exercise and even less with athletics or sports.

Combining excessive exercise with disordered eating

Many who have distorted body images that lead to eating disorders, such as *anorexia nervosas* (severe reduction in eating), *bulimia* (binge eating and then vomiting or using laxatives to purge food), or some other pattern of dis-ordered eating, begin to see exercise as a convenient — and even healthy — way to obtain their unrealistic body ideals or otherwise gain some personal control over an important aspect of their lives.

The logic here is pretty straightforward. People in general, but especially those with distorted body images, tend to think that their weight is deter-mined by simple arithmetic: If you burn more calories than you eat, you lose weight. This calories in/calories out mindset leads them to falsely (and dangerously) conclude that losing weight (their main goal) is just a matter of eating less and/or exercising more. Although this may be true in theory, body metabolism is far more complicated than adding and subtracting calories (refer to Chapter 4 for more on metabolism).

Mistakenly believing that more is always better

The numbers don't add up: Twice as much ain't always twice as good. Many people stumble into the trap of believing that, if something is good for them, then more should be even better. But that's not the way it works, especially when you're talking about the potential benefits of exercise.

Unfortunately, the official recommendations for exercise aren't very help-ful. Although most are very good at specifying how much exercise is enough, they don't seem particularly interested in addressing the question of how much exercise is too much. In fact, some organizations may even inadvertently contribute to the problem. Both the World Health Organization (WHO) and the Centers for Disease Control and Prevention (CDC), for example, indicate that healthy adults should engage in at least 150 minutes of moderate exercise or 75 minutes of intensive exercise per week. They immediately follow this minimum recommendation by unconditionally stating that doubling these times produces "even greater health benefits."

Although that particular assertion may be true, some people are inclined to erroneously extrapolate that they would get the *greatest* health benefits of all by simply doubling these numbers once again — 600 minutes of moderate or 300 minutes of intense exercise each week. At some point this "more exercise is better exercise" reasoning breaks down, as do most of the exercisers themselves.

Getting a Grip on Reality — and Exercise: Addressing the Problem

Making sure that you and the folks you work with get just the right amount of exercise isn't nearly as easy as it sounds. If you (or they) don't exercise enough, you run a huge risk of developing some pretty serious health problems, such as cardiovascular disease (heart attack and stroke), diabetes, depression, arthritis, bone degeneration, and even certain types of cancer. On the other hand, engaging in too much exercise can result in extremely painful overuse injuries (sometimes called *repetitive micro-traumas*), such as tendonitis and stress fractures; physical and psychological burnout; and detrimental impacts on relationships at home and at work.

Making things all the more difficult is the fact that everyone is different. Depending on age, physical condition, personality, and existing illnesses or injuries, what may be a perfect amount of exercise for one person would be way too much or far too little for someone else.

In Chapter 13, we discuss how to inspire yourself to begin and stick with an exercise program. Here, we tell you how to control your exercise so that your exercise doesn't control you.

Throughout this section, we talk a lot about ways to keep yourself from exercising too much. *But we are not telling you to quit exercising!* Exercise is one of the best things you can ever do for yourself. The key is to make sure you're maximizing the good things exercise and physical activity can do for you while minimizing the bad that can come from overdoing it.

Taking stock: Knowing what you hope to get from exercise

To figure out exactly how to set up healthy parameters around how much you exercise, you have to figure out what's tempting you to exercise too much in the first place. Basically, you need to ask yourself the question, "What's in it for me?"

Are you exercising just to get some of those fantastic direct (primary) positive feelings that seem to come naturally when you work up a serious sweat? Or maybe you're seeking indirect (secondary) rewards like weight loss, a better self-image, or the satisfying sense of control or accomplishment that can come from exercise.

Although there's nothing wrong with exercising for either (or a combination) of these reasons, if you ever want to limit yourself, you need to know what rewards you're asking yourself to give up and what you'll have to replace in some other way.

Thinking about what you do before you do it: Being mindful

Life can get so hectic sometimes that you may just switch on automatic pilot and go through your daily routine with as little thought as possible. Routines can help you get important things done without devoting a great deal of time and energy to thinking about them, but you need to be careful. If your exercise program has become just another thing that you do without a lot of thought, you're denying yourself the opportunity to make some interesting observations about how you can improve the way you exercise — without increasing the amount of time you spend doing it. Ask these questions:

- **Are you using exercise as a distraction?** Are you increasing your workout schedule (exercising longer or more frequently) just to avoid interacting with your family and friends?

- **When will enough ever be enough?** Can you imagine ever getting to a point where you will be completely satisfied that you're doing enough exercise and you don't have to continue trying to get a little stronger, a little thinner, a little faster, or into a little better condition?

- **What else can you do to get the same benefit?** Is there any way you can accomplish your goals without exercising quite as much? If you're exercising to relieve stress, can you take a stress management or relaxation class? If you're exercising to look better, can you improve your diet?

Committing to change

Making changes to your daily or weekly exercise schedule isn't often easy. After all, you've established this routine because you *like* it this way, right? In fact, having a routine you can count on every day often feels comforting and reassuring. But you can make a few minor — even pleasurable — modifications to minimize your chances of suffering from the negative effects excessive exercise can produce.

Acknowledging that over-exercising isn't good for you

One of the first things people who are prone to over-exercising need to do is to truly acknowledge the harm excessive exercise can cause. Simply put, you actually have to believe a real need exists to limit your physical activity, or you'll never be able to set healthy parameters around exercise. Rethinking the role exercise plays in your life may be the most difficult — but important — change you'll ever need to make.

Accepting the fact that too much exercise can be a problem is the necessary first step in any behavioral change. Acceptance and Commitment Therapy (ACT) is a popular form of psychotherapy used to treat various behavioral compulsions and dependencies. This approach maintains that participants must acknowledge that a potential problem exists before anything can be accomplished. Interestingly, the acronym ACT has also come to stand for the three steps in behavioral change:

- ✔ **A** = Acknowledgement (that a problem exists)
- ✔ **C** = Choose (to do something about the problem)
- ✔ **T** = Take action (follow through on the voluntary choice to do something or, to use the common catchphrase "git-r-done").

Adding variety

Variety is the spice of life — and physical activity, too. To vary your exercise routine, try changing the location, time, or preferred exercise activity. Doing so can make modifying your overall level of exercise a lot easier. An added benefit to changing your exercise routine is that you're less likely to suffer overuse injuries or psychological boredom. You may even find you like your new way of exercising even more than the rut you were in!

Habits are hard to break partly because they have become engrained parts of our lives. Changing even small things about the way you exercise can break your habit and weaken the hold these patterns of behavior have on you: Go to a different gym (some even honor the membership you hold elsewhere in the hopes of getting you to switch), work out before rather than after work at least one day a week, or take a spin on an exercise bike instead of using the elliptical runner once in a while.

Getting your "buzz" someplace else for a change

Many people — particularly those whose lives are filled with a lot of pressure, stress, and anxiety — seek out the positive mood changes that exercise can bring by exercising more and more. But you can get some of the same positive feelings in non-exercise–related ways. For example, substituting a yoga, meditation, or relaxation class at least once a week can help you alleviate stress and tension. Instead of automatically spending another hour in the gym pounding away on the treadmill, you can get rid of pent up anxiety and stress by treating yourself to an occasional body massage.

Setting a daily goal to do less

Typically when you think of setting exercise-related goals, raising the bar — pushing yourself toward a higher and higher level of performance — comes to mind. When over-exercising is the issue, however, you want to flip that traditional objective on its head: Now the goals is to *not* allow yourself to exceed a certain level of performance. Instead of setting a personal goal to exercise an hour a day every day this week, for example, your goal would be to exercise for 45 minutes or less four days this week.

Such a goal may sound pretty strange, but if you're not training for a competitive event, why are you pushing so hard? In fact, even if you are in the midst of training, you still need to limit yourself occasionally to give your body time to rest and recover. This type of reverse goal-setting is also absolutely essential when you're undergoing rehabilitation so that you don't hurt yourself with the erroneous "more is better" thinking.

Rewarding yourself for changing your exercise routine

Just as you can turn traditional goal-setting upside down by setting goals to exercise less, you can set up rewards for cutting back on physical activity. For example, if you really covet a pair of nifty running shoes, you can promise yourself that you will buy them only after you meet your *reduced* exercise goal for two weeks straight or after you keep your word that you'll exercise in a different place at least once a week for two weeks. Obviously, these rewards do pretty much what any set of rewards does: They give you a little incentive to do what's best for you; in this case, it's reducing or restructuring your exercise program.

Combining physical activity and social interaction

One of the major drawbacks of exercising too much is that you sometimes spend so much time engaged in a solitary activity that you end up devoting too little time to social interactions with your family and friends. Exercise and social contact don't have to be either/or propositions.

You can easily change at least some of your activities to includes others, too. Go biking, swimming, running, or for a brisk walk with someone important to you. Organize a recreational basketball, golf, or tennis league with your co-workers. Schedule a handball or squash game at lunch with your boss. Play soccer with your kids. Plenty of ways exist to keep fit while building — rather than breaking down — personal relationships with the people who are most important in your life.

At the very minimum, make sure that at least half of your weekly exercise occurs in a social environment. Enroll in a spinning class or set one day a week aside for group aerobics or Zumba. You'll stay in shape, get plenty of cardio fitness, and maybe even make a few new friends in the process.

Getting by with a little help from your friends

For many people, the suggestions in the preceding section are enough to inspire them to change their exercise routines to avoid over-exercising and the dangers that come with it. Other people, however, over-exercise as a way to deal with unresolved, deep-seated issues. For these people, reminders of the dangers of over-exercise or suggestions to be more mindful or to adopt a flexible attitude toward exercise aren't enough. In cases like these, professional help is a must.

Although not directly associated with disordered eating, exercise — specifically excessive-exercise — is merely one of the many "tools" people with eating disorders may use in their desperate attempts to take control of their lives. (Head to the earlier section "Combining excessive exercise with disordered eating" for more information on the link between disordered eating patterns and over-exercise.)

Anyone struggling with disordered eating needs to get professional help. Here's why:

- **Disordered eating is a potentially life-threatening condition:** Up to 10 percent of people who have been formally diagnosed with an eating disorder will eventually die as a result of complications related to the condition. In fact, eating disorders have the highest mortality rate of any mental illness — almost 12 times higher than the death rate of all other causes of death among females between the ages of 15 and 24 in the general population.

> ✔ **Disordered eating is a complicated mental illness:** The seriousness of the illness requires the best psychiatric and medical help possible. Some treatments shown to be effective involve individual, group, and family counseling or therapy; nutritional counseling; residential medical care; and medication. All these methods of treatment need to be closely supervised by qualified medical professionals.

Ouch! Being Harmed by Even Moderate Physical Activity

Most people with even a passing interest in kinesiology have heard of a Greek fellow by the name of Phidippides, who ran 26 miles from Marathon to Athens to tell the hometown folks about the Athenians' great victory over the Persians and to warn them to be on the lookout for retaliation. Upon reaching Athens, Phidippides dropped dead from exhaustion. What most people don't know, however, is that Phidippides ran to Athens *after* having run all the way to and from Sparta — a distance of about 140 miles *each way* — to see whether the Spartans were willing to help out in the fight. No wonder he died of exhaustion!

It's easy to see how very intense or extremely long bouts of exercise (like that of Phidippides) can lead to serious health issues. Harder to recognize is that even relatively moderate forms of physical activity, when performed by someone who has a personal risk factor, can also cause some pretty significant pain and suffering, as we explain in this section. For information on how extreme conditions can also cause health problems, head to the later section "Neither Rain nor Snow nor Heat . . . Exercising in Extreme Conditions."

Helping the heart, hurting the heart: Sudden cardiac arrest

Although regular exercise has been shown to strengthen the human heart, people who have certain underlying conditions can die while engaging in vigorous exercise. Those at the greatest risk are, as you may expect, older individuals with coronary artery disease who have lived a generally sedentary life and suddenly engage in strenuous exercise. But even well-trained athletes are sometimes susceptible to sudden cardiac arrest (SCA) while exercising, often due to undiagnosed structural abnormalities in the heart muscles, valves, or coronary arteries.

Taking steps to stop SCA

Believe it or not, in the United States alone, a young competitive athlete dies every three days or so. Many of these deaths are the result of SCA. The sheer number of athlete deaths due to SCA has caused many sport organizations to call for more complete physical exams — including a detailed heart exam using an electrocardiogram (EKG) or an echocardiogram — prior to competing.

In addition, more and more public places, like airports, shopping malls, and athletic venues, are making automated external defibrillators (AEDs) available for emergency use by the public. These portable "shocker" devices can save lives by putting a heart that has fallen out of rhythm back into a normal, blood-circulating beat.

All sport venues, gymnasiums, and exercise facilities should have one or more regularly maintained AEDs available and prominently displayed, as well as ensure that staff members receive regular training in their use.

Electrocardiograms (abbreviated EKG or ECG) and echocardiograms are two different ways of examining the workings of the beating heart.

- **Electrocardiogram:** In an electrocardiogram, a series of ten electrodes are attached to the surface of the skin at various locations on the body. The differential electrical signals picked up by these leads can give an accurate indication of the rate and regularity of the heartbeats (the waves) and a general sense of the anatomical structure of the heart.

- **Echocardiogram:** The echocardiogram produces a more complete, ultrasonic image of the beating heart. These pictures show details of the heart structures and the blood flow at various points that might indicate existing damage. One genetic (inherited) disorder, in particular, that can be detected through the use of an echocardiogram is Marfan syndrome, which tends to be more prevalent in people who are tall and thin with the long arms (wingspans) — characteristics often seen in basketball players, volleyball players, swimmers, and track athletes. The condition affects the connective tissue in the body that holds everything — including the heart and its blood supply — in place.

Exercise-induced asthma: When breathing hard makes it hard to breath

The sounds of huffing, puffing, and panting filling fitness centers, gyms, or weight rooms are usually pretty good indicators that everyone's getting a really nice cardio-respiratory workout. But for some people, this kind of huffing, puffing, and wheezing is anything but a sign that something good is happening to them.

Not surprisingly, for 80 percent to 90 percent of people who suffer from chronic asthma, exercise worsens their condition. However, even people who are not asthmatic can suffer from asthma when they engage in some pretty high-energy exercise! This condition, commonly called *exercise-induced asthma* (or *exercise-induced bronchoconstriction* [EIB] by medical types) has been around for years, but little is known about why some people have it and others don't.

That EIB happens at all seems to be related to breathing cool, dry air. When you breathe normally through your nose, the incoming air is warmed and moistened as it passes through the nooks and crannies inside the nose (called *nasal conchae* because they sort of look like an elongated conch, a type of seashell). However, when you're exercising pretty intensely, you begin to suck in great gobs of relatively cool, dry air through your mouth because the air you can draw in through your nose doesn't give you enough oxygen to run that high-powered aerobic energy system (explained in Chapter 4).

Here's the problem: Unlike your nose, your mouth has no warming and wetting chambers, so all that cool, dry air heads straight for your lungs by way of the bronchial tubes. In response to this blast of cold air, the bronchi begin to constrict, narrowing the air passage. When that happens, you're in serious breathing trouble.

Sticking to a few general guidelines can reduce the likelihood of EIB:

- ✔ **Spend extra time warming up.** Doing so results in a *refractory period,* where further vigorous exercise does not cause as many EIB symptoms. Scientists aren't quite sure why this happens, but it seems to help nonetheless.

- ✔ **Don't exercise outside on days when pollen or pollution levels are high.** Although cold, dry air can cause problems in itself, filling your bronchial tubes with all kinds of junk that's sometimes in the air (like pollen or pollution) just makes matters worse, especially if you already have asthma or allergies.

- ✔ **Don't exercise outside when it's really cold, and if you do, make sure to cover your mouth and nose with a scarf**. This traps the incoming air and warms and moisturizes it a bit before it heads down the bronchial tubes.

A great way to stay active when the weather isn't cooperating is to head to an indoor swimming pool. Exercising in the warm, moist air gives you and your bronchial tubes exactly what you're both looking for.

Repetitive-use injuries: Give it a rest, will ya?

There's hardly a person alive who hasn't suffered from an exercise-related injury. Some are major, others minor, but all can interfere with ongoing physical activity, the enjoyment of daily life, and in extreme cases, long-term health and wellness. "Weekend warriors" and well-trained athletes alike hurt themselves so much while exercising that the prevention, treatment, and rehabilitation of injuries sustained during exercise and physical activity has become one of the fastest growing, multibillion-dollar businesses throughout the world.

Exercise-related injuries really fall into two broad categories: acute macro-traumas and chronic micro-traumas. Read on for the details and to find out how to avoid these injuries.

Acute macro-traumas

Acute macro-traumas are nothing more than rapidly occurring, really big ouchies. One minute you're perfectly fine; the next, you're in a humungous amount of pain. Common injuries of this type are things like a broken arm, a sprained ankle, a pulled muscle, a torn ligament, or a concussion.

The best ways to reduce the likelihood of getting one of these acute macro-traumas include

- **Removing hazards:** Making sure all facilities and equipment are in tip-top shape before engaging in physical activity lessens the likelihood of sustaining an acute macro-trauma. Regularly inspecting playing surfaces for ruts, gouges, or slippery spots and all equipment for wear, tear, and breakage is a great place to start.

- **Changing rules or policies:** Sometimes just changing the way you play or exercise can help reduce the chances of serious injury. A great example of this are the rule changes recently instituted by the National Football League to reduce the likelihood of player concussions.

- **Using additional protective equipment:** The whole point of wearing protective gear is to keep you from getting hurt. Sometimes just adding a helmet, a set of safety goggles, a mouth guard, a facemask, or some knee and elbow pads can prevent a major injury during exercise and physical activity.

Chronic micro-traumas

Unlike the really obvious macro-trauma injuries, micro-traumas are the teeny, tiny rips and tears in the muscle or connective tissues (bones, tendons, and joints) that are hardly even noticeable at the time. Oh, your muscle or joint may feel a touch sore after working out, but it's nothing to worry about — yet.

The fact that micro-traumas aren't horrendously painful at the time is really part of the problem. After literally hundreds or thousands of repetitions of the same movements (often putting too much weight or pressure on a specific joint, muscle, or bone), the stress and strain start to take their toll, and you end up with a ton of pain that is usually diagnosed as *tendonitis* (inflammation of the tendons that attach muscles to bones), *medial tibial stress syndrome* ("shin splints" caused by repeated trauma to connective tissues that attach your muscles to your shin bone), or even *stress fractures* (tiny cracks in the bone, as we explain in Chapter 8).

Because they are often so gruesome and gory, people often pay a lot of attention when someone suffers an acute macro-trauma (all the screaming, cursing, and crying that generally accompany these injuries also make them hard to ignore). But pushing through the relatively mild annoyances caused by micro-trauma is easy — until it's too late.

These self-imposed, miniscule, internal injuries can hurt like the dickens and greatly interfere with your physical activity (they often are so painful that you can't even pick up a glass of water or comb your hair), but they're not as "sexy" as having a bone stick through your skin or having a mangled, blown-out knee. Lacking the cast, surgery, crutches, stitches, or blood that go with macro-traumas, these injuries don't seem severely debilitating at all. In fact, they tend to generate very little sympathy from others, who may believe (and you may believe this as well) that the discomfort you feel is just the kind of stuff a hard-core exerciser or athlete needs to learn to live with. The result: You fail to acknowledge the little twinges and twangs for what they are: part of your body's early warning system telling you to take a break.

Treating yourself with RICE

To alleviate the pain associated with micro-traumas, you can use an anti-inflammatory medication, such as Ibuprofen, in conjunction with the tried and true RICE treatment. RICE stands for

- ✔ **Rest:** Rest is the first, and by far the most important, element required for healing. The amount of rest required is obviously determined by the severity and location of the injury. A general rule is that, if your injury still hurts or interferes with your movement, it's not fully healed and needs some more rest and recovery time.

- ✔ **Ice:** Sometimes the inflammation (pain, heat, and redness) after the injury causes the most damage. Icing the injured area for the first 24 to 48 hours (rotating between 15 to 20 minutes on and 15 to 20 minutes off) helps relieve the pain and keeps the damage to a minimum.

- ✔ **Compression:** Swelling after an injury can cause additional pain and restrict blood flow to the area. Wrapping the injured area (if possible) with an elastic bandage keeps the swelling down without further restricting blood flow.

- ✔ **Elevation:** Another way to reduce swelling (by allowing blood to flow away from the area) is to keep the injured area elevated. This has the additional benefit of immobilizing the injury and preventing further damage.

Although many athletes and exercisers are willing to apply the ice, wrap the injured area snugly, and keep it elevated, they tend to neglect the rest component. Instead, they keep hammering away with their normal routine, and before they know it, they're sidelined with a full-blown, debilitating injury that takes months to heal.

Neither Rain nor Snow nor Heat . . . Exercising in Extreme Conditions

You've no doubt seen them, and truth be told, you may even *be* one of them — those hardy souls who push themselves through torrential downpours, blinding blizzards, subzero temperatures, scorching heat, or pitch-black darkness just to get their exercise fix for the day. Nothing seems to get in their way — and that can sometimes be a big problem. Although motivation and perseverance are certainly admirable qualities, working out when it's too cold, too hot, too wet, too icy, and so on can cause potentially life-threatening problems for those who throw caution — and common sense — to the wind.

Cold-related injuries

Engaging in prolonged physical activity in low air or water temperatures, especially when it's windy, too, can result in some really bad things happening to your body. Cold-related injuries basically fit into one of the following three general categories: *hypothermia* (dangerously low core body temperature), freezing injuries of the extremities, and non-freezing injuries of the extremities, which we explain in the next sections.

Obviously, anyone who spends too much time in very cold conditions runs the risk of sustaining cold-related injuries. Yet several personal factors can put you at even greater risk:

- ✔ **Previous cold-related injuries:** Research suggests that a person who has sustained a cold-related injury in the past is two to four times more likely to have that area of the body reinjured by the cold than someone who has never before had a cold-related injury.

- ✔ **Alcohol, nicotine, and drug use:** Nicotine constricts blood vessels and makes circulation to the extremities difficult. Alcohol not only contributes to poor decision-making that may lead people to stay in extremely cold environments longer than they should, but it can also reduce the blood glucose (sugar) level and thus interfere with shivering, which is the body's attempt to produce additional heat.

- ✔ **Body size and composition:** Both increased muscle mass and increased body fat can protect people from cold-related injuries. Women, for example, are a bit better at resisting cold-related injuries, which may be due to the fact that many females tend to have a bit more body fat to act as insulation.

- ✔ **Clothing:** The weight of the clothing isn't the only thing that makes a difference in protecting people from cold-related injuries. Wet clothing dramatically increases the likelihood of problems. Fabrics that wick moisture away from the body can keep you warmer, too.

- ✔ **Lack of caloric intake:** The body needs fuel — calories — to burn in its metabolic engine. Without proper nutrition, your body doesn't have the fuel it needs to stay warm. (This is yet another way an eating disorder may put an exerciser at risk.)

- ✔ **Illness:** A body that is already stressed or compromised in some way is likely to have greater trouble resisting extremely cold environments. People who are sick or suffer from a condition that interferes with their cardio-respiratory functioning (heart disease, asthma, bronchitis, and so on) have real trouble staying warm enough to avoid cold-related illnesses.

Non-freezing injuries of the extremities

After being exposed for an extended period of time (one to five hours) to cold, *wet* conditions (as opposed to cold, dry conditions typically associated with frostbite and frostnip), the extremities of athletes or otherwise physically active people can develop something called *chilblain* (also called *penio* or *perniosis*). When clothing or footwear gets and remains wet for an extended period of time (due to sweating or exercising in wet conditions), the exposed body parts get inflamed, turn red, and begin to itch and burn.

Hypothermia

The human body is designed to work best when the core body temperature is somewhere between 98 and 100 degrees Fahrenheit. When prolonged exposure to the cold causes your internal temperature to drop to about 95°F (35°C), many essential metabolic functions — as described in Chapter 4 — begin to operate less efficiently or shut down entirely.

Hypothermia can be mild, moderate, or severe.

Severity of Hypothermia	Core Temperature
Mild	95°–98.6°F (35°–37°C)
Moderate	90°–94°F (32°–34°C)
Severe Below	90°F (32°C)

Because you never really know a person's core temperature, even the first signs of moderate or severe hypothermia (shivering, lack of motor control, slurred speech, confusion, and apathy) need to be treated as a medical emergency. Call 911.

Freezing injuries of the extremities

When people feel really cold, they sometimes say they're freezing. Although you may feel like you're freezing, you're probably not. However, when your skin cells actually do start to freeze, you're in serious, serious trouble.

Frostbite and frostnip (a milder form of frostbite) happen, in part, because of the drop in core temperature (refer to the preceding section). When the internal temperature of the body gets so low that the body begins to shut down its life-supporting metabolic functions, a general distress alarm goes out to the parts of the body that aren't absolutely essential for survival. The message? "Send the warm blood you're using back to the body's core."

Upon receiving this message, blood vessels in the extremities (usually the toes, feet, fingers, hands, nose, ears, or face) start to constrict (narrow) so that more blood can be sent to the core. When not enough warm blood is in the extremities, the skin tissue in those areas starts to literally freeze and die.

Heat-related illnesses

Whether it's high school football players pushing themselves and being pushed by their coaches through grueling two-a-day practices in the direct sun with temperatures hovering close to 100°F (38°C) or marathon runners having to endure unanticipated temperatures approaching 90°F (32°C) without adequate water supplies, the three H's of heat-related illness — heat, humidity, and (a lack of) hydration — can combine to produce tragic consequences.

Heat-related injuries fall on a continuum from relatively mild (heat cramps) to moderate (heat exhaustion) to severe and life-threatening (heat stroke). All are the result of the body's inability to cope with strenuous activity in a hot or humid environment.

Anyone who engages in strenuous physical activity in hot, humid conditions — especially when failing to stay adequately hydrated — is susceptible to heat illnesses. But, as you may imagine, some behaviors and personal factors put some people at greater risk of developing one of these serious conditions:

- **Age:** The elderly and the young are particularly vulnerable to heat illnesses. They have a difficult time staying hydrated, and their bodies aren't very efficient in cooling off.

- **Genetics:** Some people are simply built to handle heat and humidity better than others. Those who are fortunate enough to have inherited the ability to cool off faster or more efficiently are at a somewhat reduced risk of developing heat-related illnesses.

- **Body composition and physical condition:** Those who are in excellent physical condition are much more able to fend off heat illnesses. In addition, people who are physically fit are less likely to be overweight and protected even more.

- **Heat acclamation:** If your body rarely has to cope with extreme heat and humidity, it's less likely to be able to do so efficiently. Yet, if you give your body a chance to slowly get accustomed to keeping itself cool in hot, humid conditions, it will gradually begin to adjust — at least to some degree — and be able to better protect you from heat-related illnesses.

- **Alcohol:** Alcohol is a problem for a couple of reasons. First, it's a diuretic, which simply means it makes you urinate a lot. When you drink alcohol, you may think that you're giving yourself plenty of fluids when, in fact, you're actually *removing* large amounts of fluids from your body, a situation that can lead to rapid dehydration and put you on a fast path to heat-related illness. Second, drinking alcohol can dull your senses and make you less likely to know when you've had enough exercise or heat.

- **Medications:** Drugs that narrow blood vessels *(vasoconstriction)* can cause problems because they make it more difficult for the body to cool itself. In addition, medications used to treat high blood pressure, glaucoma, or edema (water retention and swelling between the cells) are diuretics and can cause you to lose precious fluids and become easily dehydrated.

Heat cramps

Although by far the mildest form of heat-related injuries, heat cramps are excruciatingly painful and often force all physical activity to come to an immediate halt. The cause of the cramping is probably an insufficient level of potassium and sodium in the muscles due to profuse sweating.

Stretching or massaging the spasming muscle (typically in leg or foot) may provide some temporary relief from the agony, but ultimate treatment generally requires drinking large amounts of water in a salt solution. In extreme cases (or when high-level athletes need to return to action quickly), intravenous solutions can be given.

Heat exhaustion

To remain healthy, the human body must maintain a fairly constant internal temperature, somewhere in the neighborhood of 98.6°F (37°C). When you work out really hard, your body generates a lot of heat. To cool down, you sweat: As the sweat gets to the surface of your skin, it evaporates and, in doing so, cools you down.

The body is usually pretty good at cooling itself down if all it has to cope with is the heat that it generates itself. Yet when you exercise in an external environment that is very hot, your body's cooling system has to contend with both the internal and external heat, and it sometimes can't keep up. If, in addition to being hot, it is humid as well, you have even more problems because the humidity reduces the evaporation of sweat on the skin, making your cooling mechanism even less efficient. The result? Your internal temperature gets warmer and warmer.

If you don't begin to cool your body down at the first signs of heat exhaustion (cool, clammy skin; heavy sweating; dizziness; weak and rapid pulse; and muscle cramps), you're on the road to a potentially life-threatening condition known as heat stroke. For that reason, treat the first signs of heat exhaustion as a medical emergency. Cool the person down immediately and watch for any signs of heat stroke.

Heat stroke

When the internal temperature of your body rises to somewhere around 104°F (40°C), you're in big, big trouble. When your internal temperature gets this high, you're very close to dying: Your body simply can't do enough to keep you alive any longer. All essential life-supporting systems — your brain, heart, and kidneys — begin to shut down. Only immediate, proper, and extreme medical treatment can save you now, and even that's a long shot in some cases. Call 911 immediately and get the person cooled down in every way possible until help arrives (get him or her into a cool area, give ice baths, and so on.)

Part V
The Part of Tens

For a bonus Part of Tens chapter, head to www.dummies.com/extras/
kinesiology.

In this part...

- ✔ Become familiar with the behaviors and health factors that form the foundation of physical fitness
- ✔ Uncover the link between physical activity and obesity
- ✔ Discover the many careers that a background in kinesiology prepares you for

Chapter 16

Ten Foundations of Fitness

In This Chapter

▶ Engaging in activities that promote health now and in the future

▶ Using your body's metabolism and muscle-building capabilities to enhance fitness

▶ Managing calories and blood sugar levels

Sometimes the whole idea of fitness can seem very confusing. So many plans, so many ways of exercising, so many interrelated components! However, much of fitness can be narrowed to ten key concepts that, once understood, provide you with the foundation of knowledge you need to move forward with a fitness plan. Keeping it simple helps make fitness more attainable, and this chapter is designed to help you do just that.

Getting in Shape to Lose Weight

A pound of fat contains a lot of calories (about 3,500), so to get rid of the fat, you have get moving! Exercise (along with changes in diet) can be a wonderful way to burn the calories and reduce body fat.

For those who've been inactive for many years, the idea of exercise can be daunting. An unfit person, for example, may be able to walk only 20 minutes twice a week. Walking a 20-minute mile, she will walk only about two miles. Each mile is worth about 100 calories, so she'll burn an extra 200 calories per week — an amount that won't make much of a dent in the fat loss picture!

Over time, however, as she gets more fit, she can do more work and burn more calories. She may end up walking every day, and then she'll really start to burn the calories as she gets in better condition. By getting in shape, she is able to lose more weight. So give your body time to build some stamina and then just keep moving!

The nice thing about calories is that you burn them for all activity, not just exercise. Walking to the store? Calories burned! Working in the yard? Calories burned. Cleaning house? Playing with the kids? Washing the car? Same. Same. Same. So just get moving, stay consistent, and watch the fat loss go up as you become more conditioned! You can find more information on calories burning and conditioning in Chapter 4.

Making Muscle to Lose the Fat

Your resting metabolic rate, covered in detail in Chapter 4, can vary greatly. If you can increase your resting metabolic rate even a small fraction, you burn more calories every minute of the day, a situation that can result in some serious weight loss. You can increase your resting metabolic rate by doing the following:

✔ **Engaging in regular physical activity:** People who exercise consistently have a higher metabolic rate.

✔ **Adding muscle tissue:** Muscle requires a lot of energy just to stay alive. The more muscle mass you can pack on, the more calories you burn daily. To find out how to build muscle, head to Chapter 10.

Managing Calories Consumed

To avoid gaining fat, you need to match the calories you use to the calories you take in. When you eat large meals full of calories, you're eating many more calories than you need at that moment, so the excess calories are stored as fat. Changes you can make to avoid this scenario include

✔ **Changing the size of your meal:** Make it small. Have a small meal in the morning, another small one mid-morning, another at lunch, and so on. Six or seven small meals per day help spread out the calories so that your energy use matches your intake.

✔ **Including in your meal items that require effort to digest:** Digestion actually burns calories, and food high in fiber, complex carbohydrates, and some protein take longer to absorb and more calories to digest. Because fatty foods are way too easy to digest and store as fat, keep your fat intake low (maybe 25 percent of your total calories).

For more information on how calories work and what they do, head to Chapter 4.

Exercising to Build Better Bone

Strong bones are the foundation for a strong body. Although the most bone is deposited during childhood and through puberty, bone continues to be remade (broken down and built back up) through much of adulthood. Just as you can build callouses of thick skin from using a garden shovel, you can build denser bones by putting them under some stress.

Bones react to compression forces by growing stronger. Gravity and weight make excellent bone builders. Each step you take places your spinal column under a compressive load, which stimulates bone growth. If you lift weights, your bone experiences plenty of bone compression and grows. The reverse is also true: Lack of activity — sitting around, bed rest, and so on — accelerates bone loss. In Chapter 8, we tell you all you need to know about bone structure and growth.

Hormones also influence bone growth and bone loss. Estrogen and testosterone, for example, help bone growth, but they don't hang around as much in the later years, which can lead to bone loss as you age. You can offset the effects of aging on bones by staying active. Walking, jogging, and lifting weights promote the growth of strong bones and are excellent activities for all ages. So get walking, hiking, or lifting some weight. Strong bones keep a body moving!

Eating Carbs to Promote Fat Loss

Some people think they can fool their bodies into burning fat, believing that, if they deny their body calories or carbohydrates, the body will have no choice but to burn fat. Sounds good . . . except that the body has a few ways to help keep its metabolism moving. Here's why avoiding carbs won't produce the fat loss you seek:

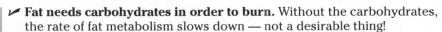

- **Fat needs carbohydrates in order to burn.** Without the carbohydrates, the rate of fat metabolism slows down — not a desirable thing!

  Carbs contribute parts to the Krebs cycle that help burn fat (refer to Chapter 4 for info on the Krebs cycle). No carbs? No fat burn. Instead fat is turned into an acidic fat called a *ketone*. Ketones are dangerous in high amounts — just ask those with diabetes who don't get their insulin and can't get carbohydrates into their cells. Ketones are so acidic they can kill cells (you included) and actually slow metabolic rate and fat loss.

- **The body's plan B for the lack of carbohydrates is to scavenge protein and use it to make glucose.** Protein is usually used for muscle building and other tissue. Using it to make carbohydrates instead is a lose-lose situation: Fat loss slows, and muscle loss increases. The upshot is that you lose muscle mass quickly — the opposite of what you want!

If you want to lose fat, your strategy should be to consume enough carbohydrates to fuel fat loss and keep the muscle you've worked so hard to obtain.

Cross-Training to Optimize Fitness

Muscles grow and adapt when they are given tasks that cause them to create new patterns of use and to adapt to tissue trauma. When you exercise the same way all the time, you're probably leaving some muscle fibers out of the action. As a result, some muscle may get strong while others stay dormant and weak, creating muscle imbalances. These imbalances can increase the risk of injury and also affect performance.

For a muscle to get the full benefit of training, it needs a changing environment of stresses. Lifting weights at different speeds and in different positions and engaging in training that involves a variety of activities help muscles continually adapt. New neural connections are made, more fibers are stimulated, and balance and coordination are enhanced. Head to Chapter 10 for a detailed discussion on your muscles.

This muscle "confusion" can really make a different in injury prevention and performance enhancement. Many athletes incorporate the concept of cross-training into their exercise regimens. They know that it fills in any gaps in their training and makes them more completely conditioned athletes. So try some weight lifting and maybe some biking. Why not swimming? Maybe some jumping drills along with some ballroom dancing. Mix it up! Your muscles will enjoy the confusion.

Accepting That Fat Goes Where It Wants

As much as we humans may try, we can't control the location of where fat gets deposited on our bodies; nor can we control where we lose fat. All we can do is control the overall amount of fat that gets deposited or lost. Here's what this information means:

- ✔ **You mobilize and utilize fat from the stores in your body in an order determined for you.** Fat comes and goes from cells based upon gender, genetics, age, and hormones (to name just a few key factors). Men may notice that fat seems to gravitate toward the belly, whereas their legs may not have as much. Women, on the other hand, may notice that their thighs seems to be happy homes for fat.

- ✔ **You can't lose fat in a particular area of your body just by exercising that area (doing sit-ups to eliminate abdominal fat, for example).** This strategy is akin to sitting in a room next to a physics class and expecting to learn physics. Instead, you need to create a demand for fat calories to be used by engaging in activities, like walking, jogging, and swimming, that use large groups of muscles (legs, back, and hips).

Keeping Blood Sugar under Control

Chronically elevated glucose does some serious damage to blood vessels and nerves, so *glycemic control,* or keeping blood sugar in a desirable range, is wise. You can attack the issue from two directions:

- ✔ **Slow the amount and rate at which glucose enters the blood:** Instead of eating the usual three big meals per day, eat about six smaller meals. And mix the foods up (fat, protein, and carbohydrate) to slow the rate of digestion. Adding fiber to the meal also helps slow the rate of glucose absorption.

- ✔ **Tackle the movement of glucose out of the blood and into cells:** Normally, the hormone insulin helps move glucose out of the blood. In people with diabetes, insulin is either not produced or it doesn't do its job. The good news is that exercise can act just like insulin and move glucose into the cells. Daily, moderate exercise both burns calories to help keep body weight down (which help the diabetes) and helps lower blood glucose. Think of exercise as the best medicine you can take for blood sugar control!

Allowing Yourself to Recover from Exercise

Improvements in fitness don't actually happen during the training activity. Instead, they happen during the recovery from the activity, when your muscles heal and grow stronger and gain the physical changes that make the next bout of training easier (head to Chapter 10 for details on muscle adaptations). Recovery is important both for the short term (so you can get back out there and train again) and the long term (improvement in fitness).

Muscles require protein (to add structure to the muscle) and carbohydrates (to fuel the muscle building process). Then they need time for the recovery and adaptation to take place. So in addition to taking time to recover, be sure to consume adequate amounts of carbohydrates and protein within the first few hours of a heavy workout. Try to go for a 4:1 carb-to-protein ratio. What's adequate? For carbs, try at least 50 grams right after the workout, which means that you would consume about 12 grams of protein to maintain the 4:1 ratio. If you skip the meal and wait too long, your recovery may be delayed for days! So work hard, but consider rest and diet as essential parts of your training. You can't improve without them!

Lifting weights daily and stressing the muscle too often interfere with the muscle's recovery and lead to reduced performance. If you don't see improvements in strength despite hard training, you may be training too much or not consuming enough calories to drive recovery.

Remaining Active as You Age

Some (young) people think that when you hit age 30, your metabolism slows down, you get fat and weak, and your life is pretty much over. As people who are well past 30, we beg to differ! Here's what the research shows: If you look only at change in fitness across different age groups, you see large declines and increases in body fat. *However,* when you look at individuals who continue to exercise, these "age-related" changes seem to disappear until people start getting to ages around 60!

Bodies need movement to keep themselves working. If you reduce your activity, your body will decline. This isn't aging; it's detraining! As years of inactivity go by, body fat grows and muscle atrophies. But this state of affairs isn't inevitable. Research has shown that even 80-year-olds can see substantial increases in muscle mass, fitness, and function when they exercise. Bottom line: No age is too old to reclaim your youth. Get moving and prove that there is life after 30!

Chapter 17

Ten Facts about Obesity and Physical Inactivity

In This Chapter

▶ Seeing what studies say about inactivity and obesity

▶ Taking advantage of your body's ability to adapt and grow fitter

▶ Finding out how to use activity to avoid obesity and the perils of being overweight

*I*n 2010, over 35 percent of adults and almost 20 percent of children in the United States were obese — a whopping total of 90 million Americans! The prevalence of obesity has grown at a staggering rate over the past 30 years, and estimates suggest that by 2050 everyone will be obese! Although that scenarios is probably not going to happen (obesity rates appear to be leveling off), obesity is clearly a global epidemic. And, boy, does it cost money!

Recent estimates put the obesity bill at about $150 billion annually, with obese individuals spending almost $1,500 more in annual medical costs due to obesity-related ailments. Although physical inactivity is not the only contributor to obesity, it certainly plays a major role, as the ten points in this chapter make clear.

As One Goes Up, the Other Goes Down

As we note in this chapter's opening paragraph and in other places throughout this book, obesity is at epidemic levels in the U.S. and in many industrialized countries. The biggest contributors to the obesity epidemic are physical inactivity and excessive caloric intake, especially involving fatty and processed foods — the hallmarks of the Western diet.

As this epidemic has grown, the amount of physical activity is at all-time lows. Less than half of adults and only three out of ten kids get the recommended amounts of daily activity. Physical education has almost been completely eliminated from school, and technology has helped create a climate of inactivity. Reversing this trend means flipping the equation: Increase physical activity and bring about drop in obesity levels.

Just Adding Activity May Not Be Enough

Just adding some activity to an otherwise sedentary lifestyle may not be enough to ward off obesity and the problems that come with it. Some studies show that when people start an exercise program, they actually reduce their activity in other areas (like leisure time) — a scenario that gets them nowhere! In addition, being sedentary for long stretches of time not only adds to the obesity problem, but it also negatively affects circulation, muscle mass, and blood sugar. Your body needs to feel gravity and movement throughout the day, and exercise may not be enough of a stimulus to counteract long-term sitting.

If you can't start an exercise program, at least stop sitting around all day! Break up work-time sitting by taking a walk to the restroom (preferably on another floor), take the stairs instead of the elevator, walk your dog before or after dinner — do anything that keeps you moving throughout the day. Who knows, you may even want to start an exercise program on top of that!

Obesity and Inactivity Can Lead to Metabolic Syndrome

As a body becomes obese, the extra fat affects a number of body systems, and the obese person ends up with cluster of conditions: high blood pressure, higher blood sugar, a greater prevalence of diabetes, abnormal cholesterol levels, and so on. Obese people also are usually quite inactive, which has its own problems.

This clustering of problems due to obesity is called *metabolic syndrome,* and it's a big reason why obesity has so many healthcare costs. Medical intervention is often necessary to control blood sugar, lower blood pressure, and improve cholesterol — all of which costs money but doesn't actually do anything to help the root cause of obesity. Long term, metabolic syndrome leads to high rates of heart disease, stroke, heart failure, kidney failure, and diabetes (to just name a few).

Reversing obesity can reverse these conditions and eliminate the need for medications. Reducing body fat to normal levels is one of the best things you can do for your body, and exercise is one of the best medicines to help you get there!

Sitting Is the New Smoking

Everyone knows that smoking is bad for you. It's linked to cancer, heart disease, asthma, and a whole host of other bad things. Even secondhand smoke is hard on the body. But did you know that physical inactivity puts you at just as much risk for heart disease, cancer, and a number of other ailments? In that way, physical inactivity is the new smoking.

Across the country, the prevalence of inactivity ranges from 10 percent to 40 percent! Now's the time to begin to view inactivity just like you view smoking and to take steps to stop it: If you have children, help them stay active and support efforts to protect (or bring back) physical education programs in schools.

The good news is that it doesn't take much activity to reduce the risks. Just get moving! Leisure activity, walking, sports, dancing, even video games that get you up and moving are beneficial. A moving body will become a healthy body if you stay with it!

Growing Up Fat Has an Emotional Impact

Kids who are overweight or obese have a much greater chance of developing deadly and debilitating diseases like diabetes, high blood pressure, stroke, heart attack — you name it. But, believe it or not, the problem is even worse than that.

Remember how awkward and self-conscious you felt whenever you didn't quite fit in with the rest of the kids in school? Remember how embarrassed you were and how much you dreaded letting anyone see even the smallest imperfection: a pimple on your nose or a bad haircut? Well, just imagine what it's like for the "fat kid" who feels awkward and self-conscious pretty much all the time.

These children are seen as fair game to be poked fun of, teased, insulted, and bullied. Such constant ridicule can have a lasting impact on the way these kids see themselves — especially when the horrible things being said about them come from the people who mean most to adolescent kids — their peers.

Compared to people who are of normal weight, obese kids (and adults) are generally seen as lazy, stupid, sloppy, dirty, and lacking self-control and discipline. That's a pretty tough way for people to go through life — especially kids — who are still trying to figure out who they really are.

Starving Yourself Just Makes You Fatter!

People interested in fat loss commonly cut way back on their calories. To some degree, limiting fat intake and the number of calories eaten is wise. However, skipping meals on the assumption that doing so will force your body to start using fat is a big mistake. What you're actually doing is setting the stage for more fat gain!

Here's what happens when you fast:

1. **Usually within a couple of days, you lose carbohydrate stores in the muscle and liver. You also lose water.**

   Because of these changes, so you may lose 4 pounds, but not of fat.

2. **Because you need glucose to metabolize fat, fat metabolism actually slows down, and ketones, which slow metabolic rate and reduce appetite, are formed.**

3. **The body searches for energy and glucose, and it uses muscle to satisfy its craving.**

   Protein in muscle is broken down in the liver to create energy and glucose. This drops more weight (but it's muscle!) and further reduces metabolic rate.

At the end of all of this, you have a reduced metabolic rate, a loss of muscle mass, and only modest fat loss. When you go back to the old ways of eating (which seems to always happen with fasting diets), you put on fat even faster! The continued cycle of fasting and then eating causes a steady gain in fat, leading to overweight and obesity. Your best option is to make wise adjustments in caloric reduction and never stop eating! You need to fuel your muscle and maintain metabolic rate. Keep exercising and reduce unnecessary fat calories.

There Is No Secret to Losing Fat

The list of diet plans and exercise programs that are currently in the media and on the market currently is enormous — and confusing! Yet if you look at all of them, you'll notice that they share a common link: limiting calories (especially calories that have no nutritional value) and increasing the burning of calories — the hallmarks of all successful diets. But you don't need a fancy diet or to spend a lot of money on a nutrition plan. Here are some general guidelines you can follow to get that first pound of fat to go away:

- **Limit your intake of calories.** Start simply by reducing portion sizes (eating on smaller plates can help). Try to eat more fiber and complex carbohydrates and drink more water; doing so helps fill you up (fat has a lot of calories without offering much tummy-filling satisfaction). Look for lower fat foods and try to avoid a lot of foods that have calories without any nutrition (think soda and high sugar foods).

- **Move.** Walk a mile, run a mile, walk the stairs, dance, play with the kids . . . anything! You burn calories when you move, and that movement doesn't have to wipe you out. In fact, if you want to lose weight, try to find an activity or combination of activities that keeps you moving for about an hour every day. This hour of activity does not have to occur all at once; it can accumulate over the day. As you begin to feel more fit, you'll notice that you take on more active pursuits, and the calories keep on burning!

This approach is a good place to start. It's simple and uncomplicated, and you can always get more complex after you've established good habits.

Fat Can Hide in Your Body

We have all pinched an inch (or more) and felt the fat that sits between our skin and the underlying muscle. But that's not the only place where fat takes hold. In fact, the fat you really need to worry about is deeper in your body. *Visceral fat* is the fat that is deep within the central part of the body (the *viscera*) and covers your organs. Here's what you need to know:

- **High levels of visceral fat, especially stored in the upper body, are linked to higher incidences of heart disease.** Upper body obesity (or an "apple" shape) is linked to greater heart disease risk.

- **Fat can migrate!** Well, sort of. The distribution of fat changes. As you age, more of your fat is stored in the visceral area — not good news. So that same inch you pinch at age 20 is not as bad as the inch you pinch at age 50.

Because obesity is so prevalent in young people, as they age, more and more of their fat will be deposited internally and create even higher risks of heart disease down the road.

✔ **Losing internal fat is the best thing you can do to improve your health and keep heart disease away.** If you are losing weight and can't see the fat melting away from your love handles, don't worry so much. It's what's on the inside that counts.

You Can Make a Difference in a Day

The wonderful thing about the human body is the way it adapts and responds quickly to physical activity. Despite years of inactivity, your body will respond almost immediately to strength and aerobic training. After only one bout of exercise, you'll experience these changes:

✔ Your body begins to adapt, growing larger muscles, building enzymes that help use fat as a fuel, and improving your ability to control blood glucose.

✔ Your ability to burn fat and control your blood sugar improves. Studies have shown that even short bursts of high intensity activity (one to three minutes) can improve control of blood sugar.

For beginners, even one day a week of weight training can add strength and muscle. If you are in poor shape, you can get in shape in a matter of months, even if you have been inactive for years! The key is taking the first step and doing something. You will soon see the results!

It All Adds Up

The good news about movement and health is that it is the expenditure of energy that reduces heart disease and cancer risk — not how hard you worked or how many miles you ran, but just the total amount of work. Studies show that if you can burn about 1,000 calories per week (equivalent to walking 10 miles, or 20,000 steps) you reduce your risk of heart disease and cancer and improve your fitness.

These calories can come from any activity. Walk the dog? Count it! Take the stairs. Go dancing. Play an active video game. Any activity that gets your body moving contributes to a healthier you. So although you may choose to work out or do some other type of structured activity, realize that as long as you are moving, you are burning calories. So just *move!*

Chapter 18

Ten Careers for Kinesiologists

In This Chapter

▶ Working with athletes at all levels and in myriad capacities

▶ Performing rehabilitative services for those recovering from illness or injury

▶ Doing behind-the-scenes work to support athletic organizations

▶ Building a career in corporate wellness

*T*he exciting part of kinesiology is the variety of careers that relate to movement. Movement is used to train athletes, rehabilitate injury, improve quality of life, and reduce the risk of chronic disease. Your career may specialize or be flexible enough to cover a variety of fields. In this chapter, we look at some of the common careers in the kinesiology field.

Cardiac Rehabilitation

Exercise is one of the best medicines available to help improve the condition of the heart and help patients regain their quality of life. As a kinesiologist working in cardiac rehabilitation, you may work with patients in the hospital immediately after the event or surgery, helping them get back on their feet and providing them with education regarding exercise, diet, and so on. Following release, you use exercise training and nutrition education and collaborate with the patients' physicians to help patients regain as much heart function as possible and hopefully prevent any future event.

One organization that can connect you to this field is the American Association of Cardiovascular and Pulmonary Rehabilitation (AACVPR) at https://www.aacvpr.org/.

Strength and Conditioning Specialist

Performance in athletics is greatly dependent upon proper conditioning. Conditioning involves a combination of aerobic, anaerobic, and strength exercises. For kinesiologists interested in using strength and power training to help improve performance, a job as a strength and conditioning specialist may be just right for you.

Often, these careers expect a certification from the National Strength and Conditioning Association (http://www.nsca.com). You'll also want to have internship or other work experience working with clients.

Corporate Wellness

Eighty percent of corporations with more than 100 employees offer some form of wellness program. Yet healthcare costs continue to spiral upward, with many of the chronic diseases (obesity, diabetes, and heart disease) leading the cost escalation. What if they had a healthier workforce? Think of the cost savings! If you like working with people one on one and enjoy spending your day conducting a variety of health-related tests, providing health information, and training clients, this may be your field!

Personal Trainer

Do you like to work one on one with clients? If so, you may want to be a personal trainer. As a personal trainer, you may work at a big sports facility, work onsite at a YMCA, or work in private studio. In any event, you'll have a variety of clients, coming in all shapes and sizes. Some may be in very poor shape, others may be training for a big event. You'll design aerobic and strength programs to help clients attain your goals. You'll also need to be able to tailor your workouts — as well as your demeanor — to their wishes.

Personal trainer certifications are available by exam through organizations such as the American College of Sports Medicine (http://www.acsm.org), American Council on Exercise (http://www.acefitness.org), and the National Strength and Conditioning Association (http://www.nsca.com).

Sport Biomechanist

Sports biomechanists examine how athletes move and look for ways to help them improve performance. As a biomechanist, you would evaluate things like how the movement is sequenced, how much strength an athlete has, or the influence that a type of equipment may have on performance (assessing the wind resistance for a road cyclist, for example). Often you'll use video taken during various performances and provide feedback to the athlete.

Check out more about this profession by visiting the websites of the following organizations: American Kinesiology Association (http://www.americankinesiology.org/featured-careers/featured-careers/careers-in-biomechanics); Texas Women's University's Biomechanics Programs (http://www.twu.edu/kinesiology/biomechanics.asp); and the American Society of Biomechanics (www.asbweb.org).

Allied Health Professions

Kinesiology covers a wide range of fields that involve movement and/or exercise as a therapy to improve or heal the body. Fields such as physical therapy, physician assistant studies, occupational therapy, therapeutic recreation, and even nursing are common career steps for the budding kinesiologist.

Often, training for these more specific health professions are graduate level programs and require strong backgrounds in anatomy, physiology, biomechanics, exercise physiology, and psychology, as well as experience designing and implementing training programs. These programs are highly selective and very competitive! You must have a very strong GPA, logged volunteer hours, and even performed some fieldwork or completed an internship in a related area just to compete.

Athletic Trainer

Athletic trainers manage and facilitate the healthcare of the physically active population. They work with high school, college, or professional athletes to prevent and treat injuries, and with physically active folks who need rehabilitative services for things like knee pain or a bum shoulder.

Athletic trainers must be certified through the Board of Certification, and preparation for this exam requires candidates to have graduated from a Commission on Accreditation of Athletic Training Education (CAATE) accredited program. State licensure is required nearly throughout the nation, and over 80 percent of Certified Athletic Trainers also possess a master's degree. For more information, take a look at the National Athletic Trainers' Association (www.nata.org) and the Board of Certification (http://www.bocatc.org).

Sport and Exercise Psychologist

Whether someone chooses to participate in some form of physical activity in the first place — and how well he or she performs once involved — is largely a matter of thoughts and emotions. Sport and exercise psychologists are academically trained to study two major questions:

✔ How do psychological factors (like motivation, anxiety, confidence, and goal setting) impact the quality of a person's performance — or whether he or she chooses to participate at all?

✔ How does the quality of a person's performance — or the fact that he or she even participates in physical activity at all — affect *psychological factors:* the way that person thinks, feels, and behaves?

To get a quick overview of what's required to be a sport and exercise psychologist, check out the web page for the Association for Applied Sport Psychology (http://www.appliedsportpsych.org/).

Coach

All sorts of coaches exist — sport performance coaches, strength and conditioning coaches, nutritional coaches, health coaches, even mental coaches. Regardless of their individual specialties, all coaches are, in some way or another, trying to improve those they're coaching. In general, the more coaches know about all aspects of kinesiology, the better they're able to help their learners accomplish physical activity-related goals.

To find out more about coaching, review the National Standards for Sport Coaches, which you can find at the National Association for Sport and Physical Education (NASPE) website: http://www.aahperd.org/naspe/standards/nationalStandards/Coachingstandardsandbenchmarks.cfm.

Athletic Administrator

Literally hundreds of behind-the-scenes jobs are needed to make any sport or athletic activity run smoothly. These jobs involve marketing, promotions, ticket sales, customer service, facilities and event management, and sports information and media relations. The head athletic administrator oversees the entire operation. In addition, he or she provides the vision and sets the overall direction for an entire athletic program.

Sport managers responsible for the day-to-day operation of a part of an athletic program usually have at least a bachelor's degree in kinesiology, business, or management. More advanced positions, such as director of athletics, generally require more advanced academic training — usually a master's degree.

Index

• T •

About the Authors

Steve Glass, PhD, FACSM: Steve is a full professor of exercise physiology at Grand Valley State University in Michigan, with over 20 years' experience in developing undergraduate programs and teaching courses in exercise physiology, exercise prescription, sports medicine, and cardiac rehabilitation. He has published work focusing on effort sense (RPE) during activity and muscle activation (EMG) during exercise and rehabilitation. He consistently links his research to student learning and all his student collaborative research has been published in peer-reviewed journals. He has worked nationally with the American College of Sports Medicine (ACSM) to establish a University Exercise Science Endorsement program, a precursor to program accreditation, and has contributed text for ACSM publications. In his spare time, Steve is an avid exerciser and weight lifter. His current passion is ballroom dancing with his lovely wife, Julie.

Brian Hatzel, PhD, AT, ATC: Brian is an associate professor in the Movement Science Department at Grand Valley State University. He serves as the internship coordinator for the athletic training major and is faculty director of the Movement Science House (Academic Community). Brian completed his undergraduate degree at the University of Florida (exercise and sports sciences with a specialization in athletic training), his Master's degree from Ball State University (exercise science — biomechanics) and his PhD degree from the University of Florida (athletic training/sports medicine with a minor in rehabilitation sciences). Brian has extensive experience in the clinical setting as an athletic trainer, providing care for athletes at the high school, professional, and Olympian level. Currently, he serves as the developer and lead consultant for a community-based throwing assessment/performance enhancement program aimed at helping athletes identify biomechanical flaws and improve performance. Brian is an active scholar in his field, and in his research, he has examined muscle firing characteristics of the glenohumeral joint, blood pressure responses during exercise, throwing athlete adaptations, glenohumeral joint arthrometry, and pedagogical/professional issues in the field of athletic training. He serves as an active consultant for many national and international performance enhancement and athletic training organizations.

Rick Albrecht, PhD: Rick is a full professor in the Department of Movement Science at Grand Valley State University. He holds advanced degrees in counseling psychology (MA) and in the psychosocial aspects of sport and physical activity (MA and PhD) from the Department of Kinesiology at Michigan State University. A recipient of the Doctoral Dissertation Award presented by Division 47 of the American Psychological Association (Sport and Exercise Psychology), he was elected to member status in the American Psychological Association in 1994. Rick is a charter member of the Association for the Advancement of Applied Sport Psychology and has served as president of the National Council for the Accreditation of Coaching Education. He also was part of the five-member team of national experts responsible for writing the second edition of the *National Standards for Sport Coaches* in 2006. Rick has worked with athletes and exercisers at all levels during the past 25 years. In 2013, he published his first book, *Coaching Myths: Fifteen Wrong Ideas in Youth Sports,* which was the culmination of the hundreds of coaching clinics and workshops he has given across the country.

Dedication

From Steve: This book is dedicated to my wife, Julie, for her endless encouragement and unconditional support. I hope to be as great as she thinks I am! I also dedicate this to my kids, Evan and Nickole, both of whom have logged a number of years hanging out in "Dad's lab" while I tried to balance work and fatherhood. You both are the pride of my life's work.

From Brian: To my most gracious and supportive wife, Gayle, and the most loving and special children ever (Alyssa, Justin, and Kayla): Thank you for supporting and encouraging me throughout this process. Even though you may not have realized it, you provided me with the focus and support I needed but often couldn't find by myself.

From Rick: To Mom and Andrea (as always).

Authors' Acknowledgments

From Steve: Thanks to Deb Feltz at Michigan State for sending the *For Dummies* folks my way. They are a team of excellent professionals who really know how to write!

From Brian: I'd like to thank my mom and dad, who taught me that nothing worthwhile is ever easy and that hard work and relationships are what define you. Thanks also to my students, who make me want to give them the best that I have to offer every day, and to the many mentors throughout my career (Kevin Mathews, Chris Patrick, Dr. Tom Kaminski, and Jim Scott, to name just a few), who took the time to teach and had the patience to stick with me as I grew up in the field.

From Rick: Like Steve, I want to thank Dr. Deb Feltz, not only for getting us started on this rewarding project but because she was my graduate school advisor, dissertation director, mentor, and friend back in the early days. If it hadn't been for Deb and Dr. Vern Seefeldt, Director Emeritus of the Institute for the Study of Youth Sports, none of this would have been possible. Finally, I'd like to thank my great colleague and friend, Dr. Dana Munk at Grand Valley State University, for her continuous support and understanding as I still try to get a handle on this professor gig, even after 25 years of practice.

From all: We'd all like to acknowledge the support and direction we received from the folks at the *For Dummies* main offices: Anam Ahmed, for getting us started; Tracy Barr, our editor, who provided constant feedback and suggestions, kept us on track, and found the right words in the right context; and Kathryn Born, our illustrator, who, through expertise and patience, created images that connect the reader to the material — an outstanding achievement considering our not-so-detailed sketches!

Publisher's Acknowledgments

Acquisitions Editor: Anam Ahmed

Editor: Tracy L. Barr

Technical Editor: Gregory Dwyer, PhD

Art Coordinator: Alicia B. South

Project Coordinator: Sheree Montgomery

Illustrator: Kathryn Born, MA

Cover Image: ©Fuse/jupiterimages